D0926721

FORTUNES
IN THE GROUND
COBALT, PORCUPINE & KIRKLAND LAKE

FORTUNES
IN THE GROUND
COBALT, PORCUPINE & KIRKLAND LAKE

Michael Barnes

 A BOSTON MILLS PRESS BOOK

Canadian Cataloguing in Publication Data

Barnes, Michael, 1934-
  Fortunes in the ground

Bibliography: p.
ISBN 0-919783-52-X

1. Gold mines and mining – Ontario – History.
2. Silver mines and mining – Ontario – History.
I. Title.

TN414.C3205 1986   622'.342'097131   C86-094779-3

©Michael Barnes, 1986.

Published in 1993 by
Stoddart Publishing Co. Limited
34 Lesmill Road
Toronto, Canada
M3B 2T6
(416) 445-3333

A BOSTON MILLS PRESS BOOK
The Boston Mills Press
132 Main Street
Erin, Ontario
N0B 1T0

Winners of the
Heritage Canada
Communications Award

American Association
for State and Local History
Award Winner

Design by John Denison
Cover Design by Gill Stead
Typeset by Lexigraf, Tottenham
Printed in Canada by Friesen Printers

The publisher gratefully acknowledges the support of The Canada Council,
Ontario Arts Council and Ontario Publishing Centre in the development
of writing and publishing in Canada.

This one is for my friend Mike Roberts
and
all those who prospect, plan for and work in the mines of Northern Ontario today.
They are a great bunch.

Contents

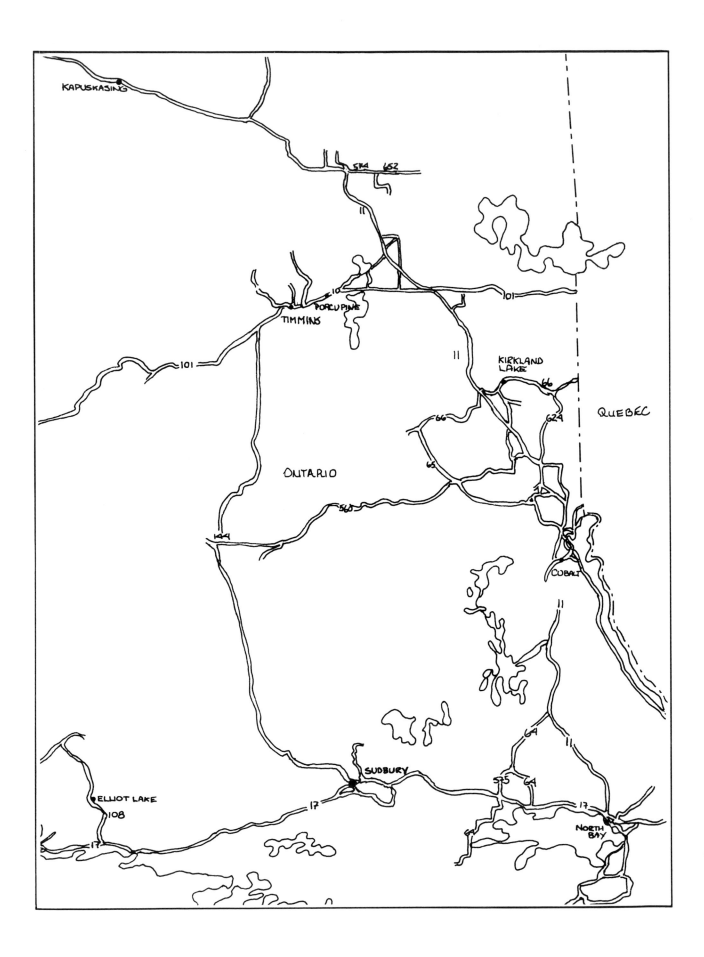

Introduction

*There are good days and there are bad days. But there is
always a challenge and a romance in mining.*
– Stephen B. Roman

Fortunes In The Ground is about gold and silver. Accounts of large sums of money
routinely appear in the book, and if any attempt is made to relate equivalents in present
purchasing power, the sums are truly enormous. This is the story not only of those who
were the big winners but also of others, losers in the games of chance called prospecting
and mining. At the beginning of the twentieth century the first of the great mining camps
touched upon in this book was found, and within eight more years two rich gold camps had
come into being. Three such areas within the span of 140 miles were to pour billions of
dollars into the Canadian economy, and all continue to operate to some degree today.

Each of the three communities that grew up around the camps were linked by
people, common ownership and material needs. All saw violent change take place upon
the land in the name of progress. Mercifully the north country is kind and the bush
swallows up and covers the worst of the blight. In some cases man's changes have created
a landscape which itself is of interest to the visitor, if not to the ecologist. As for the towns,
traces of the old idea that mining camps cannot last are still apparent in the haphazard
approach to building and town planning seen in some locations.

The author has a gold specimen given to him by a mine superintendent after a tour.
In the small chunk of white quartz and grey granite, there is a quarter-inch streak of that
warm, lustrous yellow metal. There are also slight gleams of another substance on the
other faces. These are brassy and not soft like gold. Perhaps they relate to the pyrite
family. Such samples attract the untutored eye but are not what they seem, merely all flash
and no substance.

Fortunes In The Ground should appeal to various readers. Those interested in
Northern Ontario should find a perspective of a vast, still largely unknown part of the
province. People who follow the story of Canadian mining will not only find the story of
three of the greatest mining camps in the world but perhaps some new insights into their
development. No one should make the mistake of thinking that the author is any more than
a hobbyist. His technical background counts only one geology course, a long time ago.
Maybe a lasting interest in mining is a better qualification for a book of this kind. If
industry critics find technical faults, then they are solely those of the writer.

Like its predecessor, *Link With A Lonely Land*, this book is about people. In the
north country, people make the land both interesting and worthwhile. True northerners like
to go south, but just for a visit. For many the freedom and way of life in the bush country is
reason enough for residence. In the mining towns, the man coming off shift is concerned
with the greater picture as well as the world immediately outside. He buys *The Northern
Miner* and keeps track of the industry. Miners are part of a team, their work is a skilled
job, and as a whole they keep informed about events in business which bear on their
work.

The reader unfamiliar with Northern Ontario can only get a real impression of this
great empty land by visiting the area. Take a look at the mining camps and their people, for
they have made a significant contribution to Canada.

SILVERTOWN

This is a grey wee town.
    – H.R.H. The Prince of Wales on a visit to Cobalt in 1919

Cobalt miner   – OA-S 13733

Working in a silver vein. Some of the surface veins ran for hundreds of feet.　　　– OA-18146

Proving A Camp

Where it is, there it is.   – Cornish saying, maybe related to silver

We have mined by rule of thumb long enough.
                                    – Report of Royal Commission on
                                    mineral resources in Ontario, 1891

I lost my fool self. Someone sidetracked the scenery.
                                    – Remark of newcomer to Cobalt

The newcomer to mining might well start by visiting the Ontario Legislature at Queen's Park in Toronto. On the main floor of the east wing there is a display of provincial natural resources. Right in the centre sits a huge chunk of rock, perhaps six feet long and three feet high. It is rose pink in parts. Maybe 20,000 years ago, in the last glacial age, the moving stream of ice and debris sliced it off from the country rock and dropped it near a place just about one hundred miles north of present-day North Bay. Take a closer look at that solid specimen and make sure what it looks like in case the twin ever crosses your path in the bush one sunny afternoon. The great weather-worn sample contains 9,715 ounces of native silver. No one would want to mislay such a find and likely there is more where that one came from.

French soldier Pierre de Troyes was the first white man to see minerals in the Northeast. On his way via Lake Temiskaming to attack the English forts on James Bay in 1686, the gallant adventurer was shown a mine by Indian guides on the east shore of the lake. This same showing was featured in the 1744 Carte des Lacs du Canada. It was too bad that neither soldier nor mapmaker camped on the other side of the lake. They had seen what was to become the Wright mine. Had they camped on the west bank, a much more precious metal awaited the visitors in numerous surface forms.

W.E. Logan, leading a survey party for the Geological Survey of Canada in 1884, found deposits of cobalt on the shore of the lake. The site one mile south of modern Haileybury was left untended until twenty years later, when it became the Agaunico Mine. Twelve years after the survey was made, another surveyor passed that way. Alexander Niven ran a survey line north to sort out the boundary between the districts of Algoma and Nipissing. He reported various mineral occurrences. There was the same lack of interest as in past years. Canadian lumber magnate J.R. Booth's loggers cut out white pine in the area and floated it downriver to Ottawa, but their eyes focussed on the treetops rather than the ground. That was a pity because the timber cutters ranged over ground which was rich with a precious metal so evident any passerby could have detected it.

One railway promoter referred to the Northeast as Ontario's treasure chest. The provincial government of Sir George Ross attempted to prove that statement. Ontario made overtures to the business community. There were no takers to stand the expense of building a railway. There were barely 350 people in the little community of Haileybury, near the head of the lake, and it was estimated that there were only 100,401 people in the whole of New Ontario, and that included all of the districts from Parry Sound and Muskoka through to the Manitoba border. There were no proven mineral deposits and the Booth interests had practically depleted the white pine, the only commercially valuable timber.

These boats and others like them competed successfully on Lake Temiskaming with the railway which later paralleled the lake. – Cobalt Museum

Railway construction workers like these often quit work to try their luck in Cobalt mines.

– OA-S1676

In the end the province gave in and built the line itself. The project was applauded not only by the people scattered to the north, but also by the City of North Bay, which knew well the benefits a railway could bring. The government line started north in the spring of 1902. The story of the line is found in *Link With A Lonely Land*.

One hundred miles north the pioneer line met rock muskeg lakes, and it is no exaggeration to say that it was punched through some of the most rugged country in North America. Sixty miles north it circled Temagami, and the first dividend of the line was felt when it opened up a beautiful island-studded playground as a new tourist attraction for Ontario. When the tracks reached further toward Haileybury, the work gangs came across a shimmering lake and blasted down a black overhanging cliff to form the road bed. The waters at mile 103 seemed just another lake and the line pressed on.

In common with the discovery of nickel at Sudbury and the asbestos deposits of Quebec, the precious metal at Long Lake owed its discovery to the passage of a railway. But it was only an incidental good fortune. The Temiskaming and Northern Ontario Railway was set up primarily as a colonial line. As the rails pushed forward, the main purpose was to reach the two tiny settlements at Haileybury and New Liskeard. The government road had as its further mandate the opening of the wilderness between the C.P.R. main line west and James and Hudson bays to the north.

The construction workers occasionally noticed cobalt bloom, that rather pretty rainbow on the rocks, in the spring of 1903, but no prospectors followed the tracks until mid-summer. Meanwhile railbuilders' progress was measured in track forward and there was no time in the long workdays to explore the surrounding countryside. To anyone knowledgeable in such matters, cobalt was often associated with silver, but that precious metal was not a big item in Canadian mining at the time. Only about 5 million ounces were produced in the whole country on the average per year. The three great discoveries made at Long Lake during the summer of 1903 changed the still unknown land north of the United States into a major precious metals producer.

The railway used contractors in all phases of construction. The practice was sound, as there was a very evident profit motive for on-time completion. J.J. McKinley and Ernest Darragh were tie contractors on the section north of the Montreal River. They cruised ahead of the forward-moving tracks, selecting stands of timber suitable for the very definite quality standards of railway sleepers. Teams of cutters felled the trees and hauled them to the line for treatment and shipment north. Both men had a notion to stake a claim. At first it was just an idea to break the monotony of their work. They had seen various mineral outcroppings but had no idea as to type or quality. One of the partners sent to the Ontario Department of Mines for a booklet on claim staking.

The rails had long snaked past Long Lake when McKinley and Darragh noted metallic particles in a rock cut. They leisurely followed the directions for staking and set up the required rectangle of land. The new property was registered on August 30, 1903. That claim was the first staked in the area, other than the land reserved for the railway right-of-way. Samples were chipped from the exposed rock and sent to a Montreal chemist for analysis. When later in the year word was received that the samples graded 4,000 ounces of silver to the ton, the news was as much a surprise to the two contractors as anyone else. That original claim, the JB-1, was one of the richest properties in the area, but for various reasons no mine was established on the site for three years, when backers were found from Rochester, New York. Their investment was sound, as the property amply repaid investment.

The contractors' discovery was unknown when another railway worker made his great find. Alfred Larose was a blacksmith employed by railway contractors John and Duncan McMartin. The tubby, moon-faced craftsman was bright and alert, very much interested in his surroundings. Over a period of time he noticed a pinkish-red scaling in the rocks near the tracks. It was erythrite, or cobalt bloom, which appears when the mineral has been exposed to air and becomes oxidized.

There is a fanciful legend about the blacksmith making his lucky strike by throwing a hammer at a pesky fox and noting minerals exposed when rock was chipped. Several years later, "Fred" Larose described what really precipitated his part in the litany of great discoveries. "One evening I found a float, a piece as big as my hand, with little sharp points all over it. I say nothing but come back and the next night I take pick and look for the vein. The second evening I found it. Then I go to the boss, Duncan McMartin, and say, 'Boss, I have a good thing, come with me, you give me a good show.' He say, 'Pull a gun on me if I don't.' Then I show him the vein and we stake two claims, one in his name and the other in mine." The property was recorded September 3, 1903.

The McMartin brothers and their new partner scratched away on the claims for a while and concluded that the mineral in place was probably copper. Others saw the samples later and variously concluded that the stuff was niccolite or something known as "kupfer nickel."

Fred Larose outside his black-
smith's shop.
– Ontario Ministry
Natural Resources

Later that fall Larose went on a trip to see relatives in Hull, Quebec. While waiting at Mattawa for through train connections, he took a stroll from the Canadian Pacific station to the downtown area. He bought some tobacco in the Timmins brothers' store and passed the time by casually showing the samples he carried in a cloth pouch. Henry Timmins was away on business in Montreal, but his brother Noah was most interested in the stranger's samples as he passed them through his hands. The rock was rough and had a nodular surface. It had a curious steely quality which developed quite a pleasing shine when polished. Fate had brought the right men together. The Timmins men were hard-working and ambitious beyond the ranks of north country storekeepers. They had grubstaked others in unsuccessful staking trips for the past fifteen years. Their good friend Mattawa lawyer David Dunlap had been prospecting north along the T. & N.O. line recently and remarked on the prospects in the newly opened area.

After Fred Larose put his samples back in the bag and caught the train, Noah Timmins reflected among the surroundings of the general store that his blacksmith visitor might have a valuable property. The storekeeper wrote to his brother with his urgent news. Henry was to go to Hull and seek out Larose. The letter arrived far more promptly than it would today, but the search for the blacksmith was not as fast. Henry had quite a time tracking down his man, as Larose was not an uncommon name in the city. Fred Larose finally turned up and, fortunately for the Timmins' fortunes, was in need of cash. He parted with half a share in the so far untested claim for $3,500.

The third and final great discovery of the year, 104 miles north of North Bay, was made in October 1903. Tom Hebert, a timber cruiser with the J.R. Booth Company, staked a large area on the east side of Long Lake. Rumours of Fred Larose's find suggested that the steely metal might be silver. Hebert lost no time in taking his samples to hotelkeeper Arthur Ferland in Haileybury. Ferland owned the Matabanick Hotel and was coincidentally the brother-in-law of Noah and Henry Timmins. Ferland was an astute businessman and went into partnership with two railway engineers, T. Chambers and R. Gilbraith. With Tom Hebert's claims they gathered 846 acres on the east side of the lake. The Ferland syndicate kept one parcel to the northeast, which later became known as the Chambers-Ferland Mine, and cast around for a buyer for the large portion.

Ferland had a prospect in New York. He parcelled up samples in some bags which he found lying in an otherwise empty boxcar. The mode of shipping was for security reasons. There is no record whether the recipient was dismayed when the bags marked "paving bricks" arrived. Financier Ellis P. Earle was a wealthy man, having made his money in Standard Oil. The samples told him nothing, but he lost no time in having them assayed. Promising results led to a trip to Haileybury in company with well-known mining man and promoter Captain Joseph R. Delmar. Consider the magnitude of the offer the New Yorker made to the hotelman in current funds and it is no wonder that Ferland accepted a cheque for $1 million for the 843 acres. The seal of success was set on the new mining camp, and Earle had acquired what would become the great Nipissing Mine.

Larose's original samples landed on the desk of T.W. Gibson, director of the Ontario Department of Mines, who had not been a field man for a number of years, and his initial analysis followed the popular trend. He wrote to a part-time provincial geologist and sent along the samples. Willet Miller was a professor of geology at Queen's University. He did field work for the province and skimmed through Gibson's note to its concluding paragraph: "If the deposit is of any considerable size it will be valuable on account of the high percentage of nickel which this mineral contains. I think it will be almost worth your while to pay a visit to the locality before the navigation closes."

Doctor Miller handled the sample carefully. It was obvious that the director thought the rocks contained niccolite. The geologist agreed and, even though it was October, made his way north by rail. Willet Miller was never more at home than when he was in the bush. He had a kindly face, wore a professorial straw hat when in season, and had the weathered look of a man who had seen everything. He tended to be a lot more optimistic than most civil servants of the day. The trip was just a brief one, but Miller's initial report was one of the most significant field inspections ever made in Ontario.

He noted finding ". . . pieces of native silver as big as stove lids and cannon balls." The official visitor discovered vein systems which were evident on the properties staked so far. "Four veins, all of which were very rich, had been found within sight of the railway and the fourth was a short distance to the south east. The blackened tarnished silver had up to that time attracted little attention, although it occurred in profusion in two or three of the weathered outcrops."

Miller's 1903 report of the discovery of such rich deposits of silver hit the world with all the impact of a cream puff. Many investors had been bitten by the mining bug and lost their shirts in the past. There had been fizzles before and, apart from the initial claim stakers and subsequent owners, the new silver camp was quietly ignored for some time.

More than thirty years after he became involved in the Larose claims, Noah Timmins reflected on the type of hindsight story that always abounds in mining camps. When Fred Larose and Duncan McMartin staked the original two claims, the pamphlet they relied on for directions referred to staking in surveyed territory. Under these rules, only three forty-acre claims were allowed per licence. In 1903 the Temiskaming district was unsurveyed. If the blacksmith and his boss had known, they might have secured several hundred acres and become almost overnight billionaires. Tom Hebert used the rules for unsurveyed territory and consequently was able to pick up such a large property.

Neil King enter the picture. King was a jack-of-all-trades, and in 1904 he was acting as a fire ranger when he staked a large piece of ground east of the Larose claims. He turned up at the Ottawa Valley home of industrialist M.J. O'Brien and offered to sell him the ground for $5,000. O'Brien had come across his visitor before and took one look at the shiny samples extended to him and realized that he was looking at something of value. "Why, Neil, they're not worth five thousand," he said, "but I'll give you four." That decision proved costly in the short run and rewarding over the long haul for O'Brien, as the claims overlapped and the matter went to the mining commission. The legal battle brought the McMartin and Timmins brothers together and cemented them in a solid partnership with David Dunlap, who acted as their legal counsel. Dunlap found his association with the brothers more exciting and infinitely more rewarding than, say, the $75 a year he received as retainer for acting as solicitor for the town of Mattawa. Fred Larose was fortunate that his interests were protected by the Timmins brothers. The Commissioner of Crown Lands eventually decided the claim boundaries in favour of Dunlap's clients. One happy result of such disputes was that a resident mining commissioner was installed in Haileybury.

When Willet Miller made his second trip to mile 104 in the spring of 1904, he met up with David Dunlap and the Timmins brothers at a campfire on Long Lake. As Noah Timmins recalled the chance meeting, the geologist came up with the name Cobalt as they sat before the flames. The stuff was host to silver and plentiful in the area. Someone picked up a board and used the charcoal on a burnt stick to print "To Whom This May Concern — This is Cobalt." The name stuck when shortly later railway workers and others registered at Ferland's Matabanick Hotel, five miles north at Haileybury, and wrote Cobalt as their residence. The place that was soon to become one of the most famous mining camps in the world had a name at last.

The Timmins brothers were in the area to check on their claims and do further business with Fred Larose. They optioned the rest of his claims without too much trouble and had some cabins put up on the property. At this time there were a few tents and rough shacks set in small clearings and that was the sum of Cobalt.

Noah Timmins kept a small gang of men working even though to this time he was straining the family finances to do so. There was a little silver on surface, some smaltite, (a cobalt and nickel arsenide), and not a great deal to show for unproved claims. Timmins walked around surveying the shallow trenches and sections of rock recently stripped of overburden. Noah had short legs and a somewhat rotund figure with narrow sloping shoulders but was possessed of a fierce determination to see projects through to completion. It was this attitude that made him a giant figure in Canadian mining. He may have travelled in old bush clothes like everyone else, but the man who gazed at these initial scratchings in the rock was always thinking of the next step in company development.

He recalled Miller's conversation. The provincial geologist had advised going to depth to locate promising veins. Timmins decided to sink a shaft. Work was to start on a

Prospectors at Cobalt, May 1904
– Ontario Bureau of Mines

Silver miner
– Ontario Bureau of Mines

Winter did not stop exploration.

– W.R. Forder coln.

A one horse power hoist used in the earliest days of the Cobalt camp.
– Ontario Ministry Natural Resources

Larose main vein, 1905
– Ontario Bureau of Mines

Cobbing and sacking ore, Larose Mine, November 1904
– Ontario Bureau of Mines

Friday, but one worker, a man named McCarthy, objected, as the date would be the thirteenth and considered unlucky. The protest was overcome and after some weeks the new shaft had reached the sixty-foot level. To this stage the broken ore was hauled to surface with a windlass. From his reading in the field, Timmins realized that to go deeper a drum would have to be built for cable with a tower or mast, a boom provided to swing out items hauled up, and a horse bought for motive power. He made a trip to Mattawa and on return found the work outlined had not been done. Instead the men had found a widening vein and decided to get out more ore before the owner returned.

Timmins arrived as the bucket loaded with rich ore was being hauled to surface. He noticed with horror that the bale or loop at the top of the bucket was just resting on the point of the hook. With the sway of the heavily laden bucket as it inched to surface, it was inevitable that the whole load would fall to the bottom of the shaft. As the bucket neared surface there was a creak, the hook slipped off the cable and the lot went crashing to the bottom of the shaft. One of the others ran to the pit and shouted out to the two men working below. It seemed certain that they would have been crushed, but happily neither was injured. One happened to be McCarthy, who did not hesitate to remind all in earshot of the perils of starting a job on an unlucky Friday.

The energetic provincial geologist happened by. Willet Miller predicted that these first few buckets of ore taken from depth would run to a value of $25,000 to $30,000. The money was a huge sum for the time and Noah Timmins went straight to the bank in Haileybury to secure a loan of $5,000 on the strength of the predicted value of the native silver. The mine owner cooled his heels in the area for nearly a week and then the manager informed him that the head office was not interested in mining loans. In the meantime the first two cars of ore from the new shaft had been shipped south via the railway to the American Smelting and Refinery Company, which paid prompt cash for the rich silver content. Timmins and his partners went to the bank and presented the manager with a cheque. The manager gazed at the composed features of the men in front of him and then examined the draft. "Oh, Lord," he said, " . . . William Guggenheim . . . $49,800. But gentlemen, I cannot cash that amount, I do not have the money." The men facing him across the desk broke into grins and Noah Timmins gently took the cheque out of the manager's hands. "Oh, we don't want the money," he said airily. "I just wanted to see if it was acceptable." He went on to remark that the mine owners had arranged for another bank, much more receptive to mining business, to come into the camp.

Thirty thousand dollars from that first income went to pay Fred Larose for his options. At that time the sum might have appeared large for what could still have been considered uncertain ground, but when the rich number three vein was located it was traced for 1,000 feet. The initial investment would be repaid a thousandfold in that showing alone.

W.G. Trethewey was a newcomer to the hillside scattering of tents and the odd, crude cabins of Cobalt in May of 1904. He was a Cornishman who had first settled in Edmonton and started to prosper in the real estate business. He met up with a friend, Alex Longwell, and they decided to visit the new mining camp back east. At that time there was no rush, but news of good fortune travels fast among the mining fraternity. When the two arrived on the train, they met briefly with Willet Miller, who had returned for a more in-depth examination of the area.

Miller had returned with Cyril Knight and R. Anson Cartwright as his field assistants to carry out a proper geological survey. On the second day they put a canoe into the lake — which was rapidly losing the local name Long for the newer Cobalt in common parlance. It was not long before the scientists noticed a contrast in the vicinity. The rail line was the only obvious example of progress and it was just a ribbon of steel through bush and rocks. There were the tents and a little work was observed from the shoreline, but otherwise the place was practically deserted. They studied the water's edge and noticed small arsenide veins along the east shore of the lake.

Meanwhile Trethewey and Longwell were out prospecting. They walked a short distance away from the lake, over a small hill, trundling a handcart with their packs and other gear. Longwell went off on his own, and when the Cornishman came across some promising veins near what is now Sasaginiga Lake, he prepared to stake a claim. "I had no axe with me and there were fellows down at the camp who would have made a wild rush up there if they had known. I would have lost my mine," he said later. When he did leave the camp that evening he took the axe, announcing that he was going to chop a tree down. Trethewey went back to where he had noticed the veins and moved about carefully through the bush, carefully staking the claim but at the same time avoiding contact with anyone else. That done he hiked to Haileybury to register the claim.

Trethewey vein and discovery post, May 1904
– Ontario Bureau of Mines

His companion, Alex Longwell, was not idle. He moved off to the southwest of where Trethewey staked the laid claim to a property which later became the Buffalo Mine. Almost at the same time, a prospector working for the Temiskaming and Hudson Bay Company marked out a claim just north of Trethewey's. He was working for a local firm that had been largely set up by New Liskeard merchants. Earlier that year they had found the great iron deposits at Boston Creek, south of present-day Kirkland Lake, but it was abandoned later, as there was no demand for low-grade iron ore. That deposit would stay untouched for more than sixty years.

W.G. Trethewey seemed to have the silver touch. On the same day that he staked his first property, he noticed another promising area due north of his new claim. He staked the ground adjacent to the first property, calling that second claim after his own name. He showed Willet Miller the original property, and the government geologist, after ascertaining that the ground contained copper, nickel, silver and arsenic, coined a special name for it. He used the symbols representing these principal elements to arrive at Coniagas. The resulting mine was to become one of the richest forty acres in the camp.

Miller could see now that Cobalt would develop into a major mining camp. From his observations of the lake, it was evident that water was no boundary to silver veins. As he completed his exploration trip, the provincial representative made pertinent notes and recommendations. He found that the silver was in a thousand-foot Nipissing diabase sheet which cuts across the Huronian Cobalt Series sediment and the Keewatin Volcanics, with eighty per cent of the veins showing located in the Keewatin Series. The deposits were shallow. This fact was later confirmed, for most silver was found above the 800-foot level. An exception to this general rule would be in an isolated pocket at the Beaver Mine. As at the time provincial mining law restricted claims to found minerals in place rather than "hunch" staking, Miller's recommendations that the bottoms of Cobalt and Kerr lakes be withdrawn from staking were acted upon by the province at once. The decision would prove sound for Ontario.

This miner at the Larose Mine took a well earned break.   – OA-1538032

Back on the east side of the lake, a small group of men were working on E.P. Earle's claims. They found a ridged vein, the Little vein, which was exposed from a cliff. The exact spot was brought to their attention when some of the men found silver nuggets the size of acorns lying where they had fallen from the native rock due to their own weight. The large acreage had been picked up for $1 million, and with the silver, which was quarried rather than mined, Earle had every reason to be pleased with his investment. An excavation was made "not as big as a house," and from it alone over $350,000 was recovered in just a few months. The organization of the Nipissing Mine followed this spectacular beginning.

Although the Cobalt camp had its first discoveries in 1903, the place took two years to become established and gain recognition enough to prompt a real staking rush. The only silver occurrence in Ontario had been near the Lakehead at Silver Islet at the turn of the century and it had been a disappointment to many investors. There was also the feeling among the mining fraternity and others that boom towns only developed in hard-to-reach locations, yet eventually people "mushed" to Cobalt in the comfort of Pullman cars. The final major obstacle to rapid development was the Ontario mining regulation that only areas where minerals were "found in place" could be staked. It had the effect of restricting all but rich properties. The idea was to discourage speculation and was all encompassing.

All set for blasting. Note the short fuses!
– OA-S 13752

Bags of rich silver being weighed for shipment,
July 1905 – Ontario Ministry Natural Resources

Mucking out the ore after 'firing a round'
– OA-S 13753

"A vein lode, or deposit of mineral in place, appearing at the time of discovery to be of such a nature as to contain in the part thereof exposed such kind and quantity of mineral or minerals in place, other than limestone, marble, clay, marl, peat or building stone, as to make it probable that the vein, lode or deposit was capable of being developed into a producing mine likely to be worked at a profit."

Progress on the ground was slow in 1904. Since the camp was not developed, initial operating costs were high. A box of powder sold for $32 in Haileybury and it was another dollar to transport the box to Cobalt. The rich nature of the new camp is shown in the tonnage shipped that first year of operation. One hundred and fifty-eight tons of high-grade ore was shipped south. There were only fifty-seven miners working and total wages came to $12,000. The specialized nature of the mining game is shown in the miners' origins. The first Cobalt miners hailed from Europe, South Africa, the United States and New Zealand. Their work provided $111,887 for the first investors who were beginning to take notice of the new camp.

New discoveries occurred with frequency. George Taylor's Temiskaming and Hudson Bay Company had an initial investment of $25,000 and secured one set of claims near the Trethewey property. Now prospector John Piche claimed 360 acres on behalf of the firm, just south of Cobalt and west of the tracks. He called it the Silver Queen. The Wright brothers of nearby Haileybury sold claims they had to the west of Kerr Lake and the ground eventually became the Kerr Lake Mine. Noah Timmins also turned a fast dollar on the side. He laid out $200,000 to obtain five claims near the O'Brien property and put a few men to work clearing the land. In less than thirty days he sold the lot to Bernard Barruch and the McCormick brothers of New York for three times his investment.

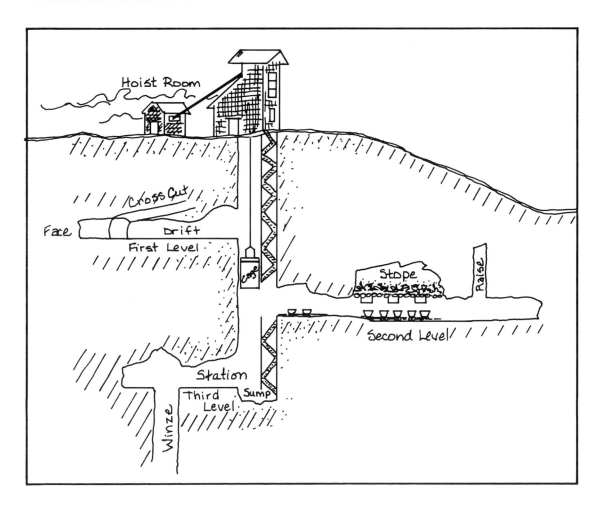

Privies are on the ridge and of these log houses at Cobalt

Animals lower floor, boarders upper. Typical early accommodations.
– OA-S 12676

Children pose on a street where most of the houses were in line.

– W.R. Forder coln.

there are no windows but one boasts a 'washing machine'.
– OA-S17202

The first Red Cross hospital was situated near the Coniagas Mine.
– Bob Atkinson coln.

Cobalt townsite was raw in 1905.
– Bob Atkinson coln.

Confidence is a prospector's middle name. Men who were tough enough to range the bush in all weather had no qualms about hanging on to a prospect in a proved camp, for buyers eventually came along. The standard forty-acre claim required a minimum of $150 per year spent on development for four years before it could be patented. Most prospectors figured that if the property was any good they would have disposed of it in short order and, if not, it could always revert back to the Crown. Prospectors Woodward, Gladden and Gates found promising ground east of Petersen Lake and held on until it was snapped up for development. They called it the Nova Scotia after their native province. Southwest of Glen Lake another threesome — Glendinning, Blair and Kerr — laid claim to the University property. One of the three, Hugh Kerr, would go on to give his name to the great Kerr Addison Mine at Virginiatown. The railway joined the prospectors in making money from Cobalt ground at the close of 1904. To complement the new station opened on December 6th, the T. & N.O. sold unserviced building lots on rough terrain which would have been ignored anywhere else. Location is all important in the real estate business, and as they were in what was now rapidly becoming the new "town" of Cobalt, the lots raked in an impressive $34,000.

By 1905 Cobalt was an established if rough-and-ready community. Picture the place as a high-top boot with the foot to the south and the railway severing the "toes" on the west, or hilly side, where most of the shacks and new houses were mixed up in a helter skelter of streets, bends and rock. This town planner's nightmare was due to the fact that the newly settled area housed an influx of men who had little interest in settling down but simply in having a roof over their heads. Building construction, after all, delayed the serious business of mining.

The province slowly woke to the fact that the men who started to arrive at mile 104 required a peacekeeper. A former Toronto policeman, George Caldbick, was made a special constable for the area. Within a year he required two assistants, and all three enjoyed remuneration of $60 a month for these positions of trust. School Inspector J.B. McDougall laid out a schoolyard while he was working to get the silver town's first school. Land was at a premium and the educator put up a sign, "School Ground — Keep Off!" Almost at once he had to get Chief Caldbick to evict squatters.

The camp was proved and growing. So far the land at Cobalt had hardly begun to reveal its riches.

First bank in Cobalt, July 1905   – Ontario Bureau of Mines

The Boom is in the Air...

Para desaroller ina mina de oro se necesita una mina de plata. (It takes a silver mine to make a gold mine.)
— Old Mexican Proverb

If a South American republic treated a European investor as we have been treated, his government would undoubtedly send a gunboat to prevent confiscation of his rights.
— Sir Henry Pellatt on the court case
Florence Lake versus Cobalt Lake (of which he was
president). The decision antagonized Premier Sir James Whitney.

In 1905 the sixteen operating mines which were just beginning to scratch the surface of the silver camp shipped $1,366,000 of high-grade ore. There were now 438 men directly involved in mining, and the ratio of miners to producing properties indicates the size of most mines. Yet if the mines were initially very small, the quality and quantity of silver found in place drew prospectors from all over the continent. Every outcropping, rock cut and natural structure was examined with the minutest of care, for any one of them might contain a king's ransom. Any one with a vein showing some calcite and silver found no lack of investors. Though the rush to Cobalt had taken two years to get under way, any southern indifference disappeared when the Buffalo, Trethewey and Coniagas mines started shipping. In some cases the pure silver from these properties was stripped from veins vertical in the rock like boards from a barn. The silver from shallow trenches was folded and hammered into convenient-sized slabs for handling. One railcar full of such specimens was sold for $34,000, and when people saw it unloaded in Toronto, the provincial capital was gripped with mining fever.

The public was intrigued by this virgin territory so convenient by overnight train from Southern Ontario. Over the next few years thousands of hopefuls participated in what can only be described as a train rush. The T. & N.O. now had a new purpose. No longer a colonization line, it carried supplies and manufactured goods. Brokers, prospectors and commercial travellers joined mine labourers on the trip north. Silver concentrates filled freight cars heading south and the provincial railway made more money from leasing its right-of-way. When in 1905 the Ontario government of Sir James Whitney cancelled many claims that were being held for speculation purposes without evident mineralization, the newcomers were further encouraged to stake claims for themselves. There was still plenty of opportunity for a man who was prepared to rough it in the bush. The new camp paid more than fifty per cent of its output in dividends, and so even the armchair adventurers could buy some shares and sit back and reap the rewards.

H.H. Lang became president of the City of Cobalt Mining Company and mayor of the town soon after. He bought forty-seven town lots and acquired mining rights to a large section of the community. It was only fitting that the mayor who made most of his mining money downtown should present Cobalt with a unique mayoral chain made of native silver nuggets as a permanent remembrance of his good fortune.

New developments in the field were commonplace. The Temiskaming and Hudson Bay Company sold the Silver Queen property at the end of Cobalt Lake for $8 million. Poet of the French-Canadian habitant farmer and woodsman, Dr. William Henry Drummond came to town in 1905. He had earlier bought a small mine at Radnot, Quebec, as an excuse to go fishing and write poetry. M.P. Wright, who was already in Drummond's employ, staked claims at the east end of Kerr Lake. The good Dr. Drummond and his family developed the ground, and the Drummond Mine became a closed corporation. William Henry Drummond was to spend the rest of his days close to Cobalt.

Habitant poet and mine owner, William Henry Drummond
– PAC-C-5363

Industrialist M.J. O'Brien survived his losing end of the Larose claim dispute with the Timmins interests and financed the development of his now legitimate ground. He needed no convincing that his original deal with Neil King had paid off when the first carload shipped south paid his private corporation $65,000. He gave close attention to the well-being of mine employees. A reporter from the *Toronto World* said that there was a bedroom for each man in a spacious bunkhouse, and a reading room, "for the miner as a rule values good reading and subscribes to several papers." With a circulating library from Toronto, good food, daily ice for drinks and ample bathing facilities, the O'Brien Mine had a good reputation in the camp. It repaid M. J. O'Brien handsomely as the longest-lived privately owned mine in the area.

The Nipissing Mine came into the news. W. B. Russell, the first T. & N.O. engineer, resigned from the railway to become a mine director. For a while there was controversy over the stock of the new mine. Major shareholder Captain Joseph Delmar had it listed on the New York exchange, but share prices faltered and he engaged William Bryce, a rising promoter, to option and push share sales. Thompson had an option to purchase shares at $3.45 each and he unloaded quite a few at that figure. Then the great Guggenheim financial house sent John Hays Hammond, their high-priced mining advisor, to look over the Cobalt property. The pronouncement was favourable and the Guggenheims agreed to purchase 400,000 shares at a sum representing an excellent profit for Thompson. Thompson was elated, but his good feelings evaporated when he found that many of his cheaply sold shares went to Delmar himself and Guggenheim had bowed out of the deal. The shares all sold eventually and made Thompson rich. Over a forty-five-year period the Nipissing Mine faithfully produced silver, with total dividends to the tune of $30 million. The share panic about the "Big Nip" was shortlived in the early days of the camp, but one disgruntled shareholder expressed his feelings in verse.

Contemplating the gifts of nature, Nipissing Mine 1907.   – Bob Atkinson coln.

31

Life is real — but this is vision
Lasting only a short time;
If you want my last decision —
To hell with NIP and Guggenheim!

From 1905 to 1907 the public interest in Cobalt spawned many bogus companies. The investment industry did not have the self-regulation that it enjoys today. The fast rise and sudden collapse of companies such as the Silver Bird, Big Ben and Cobalt Silver Mountain hindered progressive investment in the silver town. In New York the interest in silver was so volatile that at one point mounted police were used to control shareholders thronging to buy Cobalt-area stock on Wall Street. Periodic rich harvests of the precious metal were so spectacular that they fueled speculative interest. Much has been made of American involvement in Cobalt, but it was the old story of a large pool of venture capital south of the border which knew no international boundary. Canadians could have had Nipissing stock at ridiculously low prices but often hesitated, and United States-based investors took the offer. Mines like the Nipissing, Kerr Lake, Buffalo, Temiskaming, Penn-Canadian and Wettlaufer were American-controlled, but it was a case of self-interest of individuals and no organized takeover.

In 1906 the railway confirmed its new found position as a development road. An

Few spur lines to mines were very long. In the case of the Larose Mine, The T. & N.O. line ran right past the mine.
– OA-15409-4

auction of building lots netted $77,133 for the province. At an average price of less than $500, the buyer usually gained title to a rectangular parcel of craggy ground and hoped services would follow later. At this time the T. & N.O. hired Arthur Cole as its mining advisor. His handiwork may be seen in the agreements concluded that year. The Cobalt Townsite Mine paid $35,000 for its lease, plus twenty-five per cent of the value of the ore recovered. The Nancy Helen and the Wright Silver paid corresponding sums, plus a sliding scale according to the volume of ore processed. The Right-of-Way Mine straddled only sixty-six feet of the tracks yet lined the pockets of its investors as well as returning $666,915 to the railway in just eight years. For a three-year period the Larose Mine disputed ownership of silver under the tracks near its operation. The provincial railway won its case and forced the veteran producer to pay $60,000 to the T. & N.O. and its lessee, the Right-of-Way Mine.

Despite the frenzy of new building, Cobalt did not gain a hotel until February 1906. The *Canadian Mining Review* noted that building of all sorts was going on " . . . in all directions principally in the west end of town and people who have nothing else to go by but their own convenience, are building on the street and off of it." Coleman Township was set up to serve the area in the same way that municipal organization covered the town. The first sanitary inspector was appointed and made reference to Mayor Lang's angry retort to complaints that people had to buy water. "You pay for water anywhere," said the mayor. The provision of water did not prevent a town fire which destroyed sixty-five homes and another burn on the Nipissing property which did $350,000 in damage. Maybe such disasters were accounted for in high property rents. Houses valued at $1,000 had rents as high as $25 to $35 a month and insurance rates soared to as high as twenty-five per cent of property values.

Cobalt may have been growing like Topsy, but creature comforts were still relatively primitive. J.A. McRae described his early accommodation: "My first living quarters in Cobalt consisted of a little room in the Prospect House. The board walls of the two storey structure were thin and uninsulated, save for the black tarpaper which covered the outside, and which was supposed to prevent the wind from blowing through the cracks in the boards. Wood burning box stoves were the sole source of heat. In the dead of winter the water in the granite jugs froze solid, so did the contents of the earthenware pots beneath the bed. The general toilet was a lean-to shed at the rear, equipped with a two-hole stove with the lids removed. To prevent direct contact with the cold metal of the stove, a board had been carved to match up with the metal itself."

Cobalt enjoyed continued recognition within three years of discovery. One estimate during that period declared that 2,000 prospecting teams were fanning out in the surrounding bush country, while stockbrokers made frequent one-week round trips in luxurious private cars between New York and the silver centre to keep up with the latest developments. When talking to clients, the brokers likely glossed over the hazards of a camp where in one accidental blast seven tons of dynamite exploded. Neither the blast nor the resulting fire could dampen the interest of visitors or investors. Twenty mines had shipped silver to the total value of $2 million in 1906. Changes in the mining regulations helped new arrivals. The section on claims having a mineral in place was changed to read at the satisfaction of the mining recorder, but so many claims were now in existence that inspectors had to visit each one personally to check for genuine discovery. This resulted in cancellation of many speculative properties and freed up more ground for exploration. The *New York Times* in May 1906 reported on the hopefuls who tried their hands at prospecting in newly opened areas. "Some old-time California placer miners and there are others fresh from the copper mines of Arizona and Senora. Not a few are from British Columbia and the Yukon camps. All there rub shoulders with the Kimberley diamond digger and his brother from the mining fields of British East Africa. The boom is in the air and it's contagious . . ."

COBALT'S FAMOUS SILVER SIDEWALK

Such a source of nature's bounty was a constant excuse for reflection. This site is easy to find today but the 'sidewalk' is long gone. – Cobalt Museum

Life was not all glamour in the camp. There were few labour-saving devices and a miner's lot was long and hard. A few took solace in taking their work home in the form of native silver highgraded in lunch pails and pants pockets. The newly formed Mine Owners Association was able to secure convictions against six who tried to share the wealth in this manner. Labour was still scarce and there was a general call for union activity to obtain higher wages and standards of living. Although daily pay included room and board, it was for a grueling ten-hour day. Machine and hammer men made $2.25 for their ten-hour stint, underground miners $1.75 and surface employees took home $1.25.

Cobalt confounded the rule by remaining a low investment camp. Despite the rich ore, in 1906 the total expended in machinery was likely not more than $100,000. Wheelbarrows, picks, shovels, drill steel and hammers were still the basic work tools. No one saw the need for protective gear such as hard hats. The fabulous silver output still came from a handful of small headframes, which would be dwarfed in later years by those of the great gold camps to the north. Even today it is easy to see how close the rich deposits were to the surface. The cuts in cliffs and tortuous cracks in the rocks, sometimes snaking along for hundreds of feet, were made by pure physical activity, as pick and shovel men were aided only by the odd dynamite stick. R.J. Barrett, a Canadian financier, tried to explain the wonder of a place where silver came in fissures, veins, water, "sidewalks" and even "waterfalls." "At Cobalt it seems almost impossible to exaggerate; the riches were so palpable there; any urchin in the street could unerringly take the stranger to half a dozen mines and point to wonders which fairly beggar description."

Gold was discovered in the Larder Lake area in 1906 and some Cobalters were lured away to try their luck as prospectors. The Crown Reserve Mine at Kerr Lake started operations in the newly opened area, but the gold showings appeared sparse and the 6,000 men camped in the bush quietly folded their camps and slipped away. Nearer to Cobalt, there were some silver finds at Casey Township, but nothing to detract from the boom at Cobalt.

More than $3.5 million worth of silver was shipped from Cobalt that year and anyone who was not aware of the camp at mile 104 did not read the newspapers. Over the next two years, 300 mining companies did business in and around Cobalt. But caveat emptor, only twenty-two ever shipped ore. As an illustration of how money becomes concentrated, seventeen of these shippers actually paid dividends, and of this number, seven were bought and sold by the Timmins interests.

Water was ever a factor in any mining decision made in the Cobalt camp. Mines like the McKinley-Darragh, for example, had to watch Cobalt Lake very closely in relation to shoreline workings. As mines went underground, seepage was always a problem. Silver veins have no respect for boundaries and often plunged under lake bottoms. Thus when the province finally decided to allow Cobalt and Kerr Lakes to be mined, interest ran high in the mining fraternity. The Crown Reserve Mine took the rights of Kerr Lake for a lump sum plus ten per cent of gross receipts. Sir Henry Pellatt, wealthy stockbroker and builder of Toronto's Casa Loma, was president of the Cobalt Lake Mining Company. His firm paid the previously unheard of fee of $1,850,000 for the privilege of mining the forty-seven acres of Cobalt Lake. Over the next six years, six shafts were sunk to intercept veins from the Larose and McKinley-Darragh mines. A vein system was discovered which in some places ran to ten feet wide and carried silver values of 4,206 ounces to the ton. Two ore cars of this high-grade alone brought $30,000 when sold.

Annoyance with government meddling in the private sector is as common today as when Ontario entered into the mining business for itself. Extending for about one hundred miles on both sides of the Montreal River lay a rich concentration of red and white pine forest known as the Gillies Limit. The province had closed the area to prospecting, fearing that valuable timber would be destroyed when exploration work took place. This prudent step had the unfortunate side effect of persuading the mining community that Ontario was holding on to rich silver ground. Rumours abounded. One rumour had it that the sale price for the Limit would eventually be $20 million. Several told of prospectors who had secretly entered the forbidden territory and found rich veins. As the stories persisted, the province eventually offered bonuses to anyone who could prove the claims of rich silver in place. In some cases the so-called prospectors actually led geologists into the area, but always seemed lost when it came to locating a "find." Eventually an exploration party was sent in who located a seven-inch vein carrying smaltite and niccolite. Accompanying shoots of native silver and nuggets prompted further investigation, but the vein petered out and Ontario's venture into mining came to an end. When the Gillies Limit was finally opened to staking, it was found that the ground was better left to the lumber industry.

Pause and look at the town that silver made. Miners came from all over the world. Some from places like the Yukon and Alaska had seen the end of their eldorados and came to Northern Ontario in the hope of regaining their dreams. They found a busy boom town which was surprisingly law-abiding. The newcomers saw roads built to avoid high rocks, steep gullies and the ever present headframes and mills. Cobalt theoretically had been laid out on contour maps to give the most convenient situation to its streets, but there was so little space leading down to the lake that instead the thoroughfares were laid parallel and at right angles to mine boundaries. The place was no beauty spot. Mine managers and owners felt they needed more than just rugged charm and built grand homes at Haileybury facing the lake. By contrast Cobalt accommodation for a long time consisted of rough boarding houses with single rooms and facilities out back. Today there seems such an incongruity between a town built among such riches of nature and its appalling sanitation. There were no licenced premises in Cobalt, but three banks and two churches served other human needs. An opera house was to be built and the new Y.M.C.A. would have the first swimming pool housed by that organization north of Toronto. With six-day work weeks common, the new workers barely had time to sample civilized amenities. One shift worked while another made use of the warm beds they had left.

The gold rush to Elk Lake and Gogama in 1907 drained some miners from the silver town, but they were usually replaced fairly soon, often with recent immigrants. Prospectors continued to look for new "prospects." Such work had to be more systematic now, for the days of snow staking or blanketing ground were gone, as so much land was staked. The top fifteen producers were Coniagas, Nipissing, O'Brien, Buffalo, Trethewey, Larose, Silver Queen, Kerr Lake Foster, Temiskaming and Hudson Bay, Green-Meehan, McKinley-Darragh, Nova Scotia, Townsite and the Right-of-Way mines, and between them they accounted for over ninety per cent of all worthwhile ore. The Nipissing was recognized as the greatest producer. Willet Miller's map shows the company as largest groundholder, completely surrounding Peterson and Carr lakes and fronting almost all of the east side of Cobalt Lake. Happy were the Nip's shareholders. On the average, sixty per cent of its earnings lined their pockets.

Public interest in the silver camp never waned and news of Cobalt was sure of a spot in southern newspapers almost continually for several years. There was ever some striking new vein or development to encourage reporters. The Right-of-Way Mine produced the first air hammer to be used in the camp. The location of the silver deposits required ingenuity on the part of the mines. Rail transport costs were high to southern Canadian and United States refineries. The McKinley-Darragh Mine built the first stamp mill, and by 1907 nine such plants graced the skyline close to town. The problem in processing the local ore was in the host minerals. The silver was joined by arsenic, cobalt and nickel. Metallurgists and engineers from all over the continent worked on silver extraction. Cobalt was discarded as just a nuisance, but fifty years later men would comb the old ore dumps for the silvery-white, lustrous, magnetic element.

Ice cream parlour, tobacconist and lawyer all shared the building with the government assay office, 1907.

– PA-56054

At first ore was sorted, bagged and then shipped out. Anything less than one hundred ounces of silver to the ton was considered too expensive to ship. With the advent of stamp mills, the process was speeded up. First the ore was broken at surface by a gyratory crusher into one- or two-inch lumps. Then it was sprayed with water to allow hand picking of highgrade ore. What remained was trammed over to a gravity crusher. Here the ore was passed through ball and roll crushers. After sorting on screens it was pounded by cast iron stamps. A long, upright steel stem with a shoe at the lower end weighing 1,300 pounds was lifted almost seven inches by a cam on a revolving horizontal shaft. This was then allowed to drop on a die, another heavy casting, at a rate of about one hundred times a minute. The stamps pulverized the material effectively, but their roar was heard all over the area. The concentrated residues were dried and shipped south to be refined. Unfortunately several ounces to the ton were missed by these methods and it was the need for refinement which occupied the mines for a long time.

Fred Latchford, who acted for the province at the opening of the railway, was involved with the Hunter Cobalt Mine. Bob Bryce managed first the Silver Queen and then the Beaver mines. Within a few short years he would go north to start other mines in a new camp. Clem Foster (of the mine of the same name) became mayor of Haileybury, and he too later went north to bring in a new mine. All of these men were overshadowed by one man in 1907.

Dr. William Henry Drummond and his brothers had in their mine on the east side of Kerr Lake a small but profitable concern which, since it was privately owned, was not obliged to report profits or other vital statistics. Drummond lived on a hill in Coleman Township overlooking the mine site and enjoyed his life as mine owner, fisherman and still-active poet of the habitant life. A friend was fleeced by shady prospectors and Drummond wrote a poem, "The Calcite Vein," about a worthless limestone and chalk showing.

So it ain't ver long w'en we mak a strike,
W'at d'ere calling de vein calcite;
Quarter an inch . . . jus' a little "pinch" . . .

There was a smallpox epidemic in Cobalt and Drummond could not forget his training. He helped a local doctor treat patients and soon became overtired himself. He suffered a stroke, which led to a coma, and he finally died at his hilltop home. Drummond died as the boom was in full force in the silver town. His poems had passed into a permanent place in Canadian poetry. Sir Wilfred Laurier unveiled a memorial to him in 1911.

Cobalt enjoyed a media boom in the United States when newspapers picked up an article about a sample of rich ore from the Trethewey Mine which was two thirds silver. But mile 104 was not the only place where the precious metal was being found. A farmer named Bucknell saw samples of Cobalt bloom and then remembered where he had seen it before. He kept his observations to himself and went home, fifteen miles north to Casey Township, near present Belle Vallee, and staked what became the Casey Cobalt Mine. The property produced $2 million worth of silver, and its name was just fine because anything with the name Cobalt caused investors to sit up and take notice.

Years before, one of Booth's timber cruisers was working in the South Lorrain area, twenty-five miles southeast of Cobalt. He noticed native silver and blazed a tree at the spot, but since he was employed to pick out commercial timber, kept on with the original task. Later he was unable to return to the spot, but prospector Charlie Shields came across it. That property was acquired by a Doctor Wetlauffer and took his name. The area became known as Silver Centre and was host to several mines. While never as big as Cobalt, two of the mines that were developed there, the Keeley and the Frontier, were fine producers. Robert Jowsey, Charles Keeley and John Woods found silver there when they lit a fire to thaw out the ground. Both Jowsey and Keeley were to be involved in another mining camp to the north.

The South Lorrain mines; Frontier (left) and Keeley hardly show in the surrounding bush but produced fabulous silver.

– OA-S17940

39

The South Lorrain camp provided several spectacular chunks of silver. The Frontier Mine retrieved one huge specimen, which was obtained for display at the Royal Ontario Museum. The Keeley "nugget" was perhaps more famous. It weighed 4,402 pounds and contained 24,222 ounces of silver. The specimen was exhibited at Wembley Stadium in England and finally found a home with the Ontario Legislature at Queen's Park.

Outside properties could not eclipse the special nature of Cobalt. The silver town was incorporated in 1907 as the first town in Northeastern Ontario. Evidence of its new respectability could be found in advertisements in the *Cobalt Daily Nugget*: Temiskaming Steam Laundry . . . "Why have your wife, daughter or sister make her life a burden during the hot summer days, bending over the wash tub when the Temiskaming Steam Laundry is in such a splendid position to do all this slavish work at so reasonable a price and with such quickness?" There was even a social strata in the boom town, culminating in the Cobalt Mess, a popular rendezvous for mine officials.

Pioneer Ontario Provincial Policeman, George Caldbick.

Provincial policeman George Caldbick saw colleagues appointed to support him. In addition to the peace officers in Cobalt, the province sent a policeman for Haileybury and one for Charlton to take care of problems caused by the influx of railway workers. Caldbick had an interesting job and he "maintained the right" well. Cobalt may have been mostly just boisterous, but occasionally potentially ugly situations did develop. Once a character from the West announced his intention to shoot up the town. Guests at the Cobalt Hotel, just opposite the station, were lined up against the wall while he informed them of his plans. Caldbick just happened to come in and stepped forward to greet the stranger. He was told to get out fast. "Let's shake hands before I go," said the officer. Perhaps he squeezed too hard, but the man went down on his knees and the gun went flying. Handcuffs were clapped on the miscreant and life settled back to normal. As a matter of policy, Caldbick met trains and searched for guns, so serious problems were rare in the silver-rich community.

One big Cobalt "name" did not reside there but visited periodically to transact business. M.J. O'Brien maintained his home base in Renfrew. The millionaire industrialist did not interfere in the day-to-day operation of his successful mine. Usually when he visited the silver town, he remained in his private railcar and conducted business there. Despite the early claim dispute with the Timmins interests, his property had shipped $1 million worth of silver by 1907. The press loved his manner and referred to him as a bearded patriarch. He was a mine owner, manufacturer, philanthropist and nation builder, for his firm constructed a large part of the National Transcontinental Railway, now the C.N.R. The big, energetic man — who one source indicates just bulled his way through under the best of advice — became annoyed with processing choices for his concentrate and bought the Deloro Smelter and Refinery of Marmora, Ontario, to cover that end of the bullion recovery. His mine made him between $7 million and $8 million — but this is an estimate, since the business did not report profits.

The T. & N.O. started its Cobalt Flyer to serve the special needs of the camp. The train left Toronto at nine in the evening and arrived in Cobalt around the same time the next morning. The *Special*, the more prosaic name the railway gave to its crack train, offered three business cars, three second-class coaches, two firsts and three Pullman cars. In the upper-crust end the wealthy dined, and in second the humbler folks travelled on their way to work. In his *Sunshine Sketches of a Little Town*, Stephen Leacock recalled seeing such trains en route to Cobalt. "On a winter evening . . . you will see the long row of Pullmans and diners of the night express going north to the mining country, the windows flashing with brilliant light, and within them a vista of cut glass and snow-white table linen, smiling negroes and millionaires with napkins at their chins swirling past in a driving snow storm." As Toronto was blessed as the place where one changed trains for Cobalt, the railway itself smarted under local jibes. The *Cobalt Daily Nugget* complained " . . . if the T. & N.O. deserves to be called the people's railway, they should put on more local trains and treat the silver belt as a private corporation would." This criticism was countered within a few years by another form of train service.

There were more serious problems than train service in Cobalt. It was inevitable that labour unrest should develop. Hours were long, safety precautions almost nonexistent and pay was not high when mine profits were considered. The Cobalt Miners' Union wanted a six-day work week of not more than nine hours a day. Those fifty-four hours could become standard, but they also looked for three shift periods to be set on a fixed basis rather than varying from mine to mine. The pay sought was from $3.25 to $3.50 a day. Mine owners became alarmed and locked up dynamite supplies, but there was no cause for concern. The strike was short-lived, but some men made the best of a bad thing while out of work. They picked spruce gum, which was used by southern manufacturers to make the "amber" in men's pipes. The City of Cobalt mine was the only one to accept the terms sought by the union. Other mines waited out the men and several brought in outside workers. By December 17th the strike was over. The miners gained nothing except the promise of bonus pay. The issue had merely been shelved and would not go away.

A popular figure with a homburg hat, downturned mustache and generally rakish air visited the town he had named. Willet Miller saw Cobalt five years after his first visit as a place that announced itself in statistics guaranteed to amaze. Sixteen and a half million dollars worth of silver was shipped that year. The mine work force of 2,500 men produced dividends that beckoned the investing public. The Nipissing Mine recovered $1,504,833 of silver-made dollars. The Larose brought in $43 every hour or a cool $1,000 a day.

The precious metal mined at Cobalt was only exceeded by the silver mines of Mexico. The fortune in the ground was exciting just to extract, for it came in such pleasing variety. Some was like feathery snow, other finds appeared like a film of polished metal. Arborescent silver was like a frosted tree. That spun like a web, found in calcite veins, was called wire silver, while a pretty tinted variety delighted in the name ruby. Some was found in plates or sheets, while other deposits were like leaves in the fissures or even waterfalls. The famous silver sidewalk of the Lawson Mine had been planed down by the glaciers and dipped under Kerr Lake. Picture sterling silver in a vein sixteen inches wide. Passersby treated the outcropping as a good luck charm and scraped their boots on the stuff to keep an ever present shine. Over at the Nova Scotia the jewellery came in plate size. The Kerr Lake had chunks of native silver encrusted with ruby silver, which ran from black to ruby red. Nature went overboard at Cobalt.

Prospectors must have felt the precious metal had an affinit for lakes around Cobalt. Cross Lake, north and angling southeast of town, had seven mines. The Colonial, Violet, Watts (or King Edward), Victoria, Temiskaming, Lumsden, Beaver and Rochester were either on its shores or nearby. There was so much activity that a small steamer ferried people around the lake. Similarly, Glen Lake to the southeast was a centre for the Cobalt Central, Bailey, University and Foster mines. At Kerr Lake the waters were a focal point for the Drummond, Kerr Lake, Silver Leaf and Lawson mines. The latter, like many other Cobalt properties, had disputes over ownership and was finally in good standing when Larose president John McMartin purchased the property in 1908. Colonel J. Carson and the syndicate that obtained the rights to Kerr Lake for $178,000 easily promoted working capital, and the Carson vein of the newly created Crown Reserve Mine provided 20 million ounces of silver all above the 200-foot level.

Wages were inching up as profits soared. Standard Cobalt pay as of July 1, 1908, for a shift of nine hours, was $3.50 for machine men, while surface labour received $2.50. Miners took home $3, hoistmen $3.10, with the highest rate for head blacksmiths at $4. They worked in a camp which sought technology and progress. While seventeen diamond drills dotted around the camp averaged 6,000 feet a month, of eighty properties only forty had air compressors. The problem with compressed air was that it extinguished the candles

The lake in 1910 was polluted with mine wastes but the launch ride beat the state of the roads.
– Cobalt Museum

42

used for underground lighting. Mines tried various types of lamps using oil or acetylene, but they were all unreliable or needed cleaning underground. While the industry wrestled with these problems, gelignite was introduced as an explosive agent. Unfortunately it became sensitive when frozen and was not retained for long.

As the camp grew by up to one hundred newcomers a month, the railway completed a close to four-miles-long spur line to Kerr Lake to serve the mines there. T. & N.O. rates for freight were heavy and occasional shortages of boxcars caused mines to stagger shipments. These annoyances made headlines in the breathless prose employed by the *Cobalt Daily Nugget*, which found still more excitement in criminal matters.

Cobalt must have been a disappointment to those who would like to equate the silver camp with frontier towns of the American West. Liquor offences accounted for most police time. There were hotels in New Liskeard and Haileybury, but forty blind pigs or illegal public houses flourished in Cobalt, and Caldbick and his police colleagues managed to obtain convictions against seventeen of them, with a total of $1,750 in fines meted out. The bootleggers paid their fines and kept on supplying suds. Next to claim jumpers, the most heinous crime in the lexicon of honest miners was highgrading, but it was inevitable in a rich camp. In 1908 the official estimate of precious metal spirited out of mines in clothes or lunch pails was $32,000. A jeweller was arrested for possession of stolen property in the form of native silver and a Nipissing miner received an eighteen-month jail sentence, but the practice was never completely stopped.

There was no land to waste on some properties.   – OA-15409-11

George Caldbick's big case was a major theft from the Nova Scotia Mine. The matter was serious, for high crimes were unknown in Cobalt. The mining company complained that the strongroom door had been forced and twenty-five silver bars stolen. Caldbick checked the crime scene carefully and found twenty-*six* bars missing. The provincial policeman and town officers searched for a considerable time and finally located the missing bullion in a dried-up creek bed. An anonymous tip eventually led to the arrest of four miners. Magistrate Siegfried Atkinson, a noted northern character in his own right, remarked, " . . . at least our northern criminals are of a superior type, not the nasty pimply faced kind that they have down in Toronto." He reflected that the loot had been recovered and that it had not even been well hidden. Then it turned out that the stolen silver had been taken on a bet without real intention of theft. Justice appeared satisfied with the accused placed on probation and liable for the damages to the mine strongroom.

By 1909 the collective output of silver from Cobalt had exceeded the value of the famous 1898 Klondike gold rush. Even at only fifty cents an ounce, local silver brought $12,531,425 in one year. But even in this rich camp the surface finds were largely gone and the seventeen diamond drills worked steadily exploring mineral deposits at depth. Mining companies reorganized to make the best of their resources. The Larose took on the Extension, Princess, Silver Hill, Fisher, University, Violet and Lawson mines. The Peterson Lake Company covered its assets by leasing 155 of its water-covered acres to twenty-nine syndicates for twenty-five per cent of the ore recovered from each property. The Right-of-Way Mine was limited in lateral expansion, so it leased the four more miles of track right of way. At this time th provincial government cut its losses and auctioned off the Gillies Limit. The profit on the sale of mineral rights on the property was all that land gave up from mining.

The quest for technological advance saw the Nipissing Mine first use one hundred men to cut over thirty-three miles of trenches and then use hydraulic monitors to jet water over the exposed rock to expose all surface veins. No wonder the land east of Cobalt Lake is now so barren and scarred. A photograph of the time shows Fred Larose with a bowler hat, bow tie and a prosperous look. The former blacksmith could remember when he made his famous discovery in an area now dotted with headframes. Steam hoists replaced hand cranes, diamond drilling was now conducted with pressure drills and all the producing mines were at depth. The Larose dropped to 325 feet, the O'Brien to 300 feet and the Nipissing to 200 feet. Great mills were built down sloping hillsides to take advantage of gravity flow.

The cost of shipping prompted activity on the part of the major mines. The Buffalo Mine built the first flotation concentrator in 1907 and similar plants followed. The Nipissing Mine made a contribution in refining processes. In 1909 the O'Brien mill made a breakthrough in cyanide milling and the days of profitable ore shipments out of Cobalt were numbered for the government railway.

At the O'Brien mill the crushed rock was reduced to slime until the minute silver particles came free. Sodium cyanide was then used to treat the sludge in huge mixing tanks. Aluminum dust came next in the circuit, as it was used in the cyanide solution to precipitate the silver. The end product was melted down and poured into bars. M.J. O'Brien turned down offers of $4 million for his mine once the cyanide process was perfected. His normal business skills were still very much in evidence about the same time, as he bought a new property for $10,500 and sold it for a rumoured $400,000 in a couple of months. On one of his trips to Cobalt, the mine owner-industrialist chuckled to himself when a chance companion in the smoking car unwittingly talked about his casual acquaintance. "There's an old bugger who owns a whole mine by himself and every penny that comes out of it."

O'Brien probably never met Roza Brown on the train, as he was in the Pullman car

Mines on the hillsides were facilitated in their gravity flow operation. The Cobalt Mining Company, 1910.
                                                                              – OA-S18656

and she was cooking, but that was his misfortune. Roza Brown was Hungarian. She had emigrated to Canada with her husband, a former army officer. Something led her to Cobalt and she opened a bakery business there. She made the dough and then went home, while her husband kept the fire going and looked after the shop. Trouble was that Brown was a drinker. One morning Roza came back to find her husband dead drunk and asleep in the rising dough. She soundly berated him, employing the end of her broom for good measure, and used the dough anyway. No one remarked on the bread, which might have been darker than usual that day, but at that time the Browns went their individual ways. The husband disappeared from northern history, but Roza took a job cooking on the railway until she became very much involved in another mining camp — and that story comes later.

Power preoccupied all mine owners. Electricity costs were so high that one drill, operated eighteen hours a day, could run to $250 a month. Coal was $5.50 a ton, or way more than a miner's daily wage, yet the black gold was a necessary part of life and more than 100,000 tons of it were used in 1909. In a town where a fluctuation in price of only a few cents per ounce could close a mine, the new compressed air plant being built at Ragged Chutes was seen as a boon to cost-conscious properties.

There was another downtown fire, this time on July 1, 1909. While the newly formed Ontario Provincial Police aided in cleanup and security after the burn, the *Cobalt Daily Nugget* as usual put a good light on the catastrophe. The main headline promised "A Greater Cobalt to Rise Phoenix-Like from the Ashes." The fire did not damage the recently completed second Cobalt railway station. (The 148-foot by 31-foot building still stands.) Researchers will find conflicting news from the time. One source has it that none of the streets were "worked." Another version, published in booklet form by the *Canadian Mining Journal*, ignored what some said was the noxious town incinerator and malodorous garbage pile. In a summary section for investors it offered a glowing report: "The

Completing the second Cobalt station 1909. The rugged nature of the mining town is shown. The station closed June 30, 1983.
                                                                                    – PAC-ss953

Township of Coleman expended thirty thousand dollars . . . on roads to connect the outlying mines with the town. In the town itself, good streets and sidewalks have superceded trails, a modern lighting system has been installed and a water and sewer system involving expenditures of approximately $100,000 is under way. While Cobalt will never be a beautiful or even attractive city, it has gradually developed into a town as active, as well organized and well governed as any municipality in the Province of Ontario."

News filtered down the line of a large gold discovery south and west of Cochrane at a place called Porcupine Lake and, another closer to home at Gogama. Inevitably some locals left to try their luck, but the silver town noticed no real loss. Noah Timmins reflected later that all those intimately connected with the great Hollinger mine to the north came together in the Cobalt camp. Both they and others could be counted in the select thirty-five who at last count had been millionaires due to their gains in the silver town. The *Cobalt Daily Nugget* proclaimed with its usual rich style " . . . here the ground breeds millionaires. One bumps into them everywhere. Yet only a year or two ago, many of them were labourers. The venturesome and rugged weatherbeaten prospectors quickly followed, and in five years since the region has been turning out millionaires almost as fast as the nuggets."

In 1909 the Florence Mining Company took action against the Cobalt Lake Mining Company in the Ontario Court of Appeal. The Florence grievance was that for the previous two years the Cobalt Lake Company under Sir Henry Pellatt had held the rights to the area covered by Cobalt Lake. The appeal seemed odd because everyone knew that Pellatt's group had received the rights to the lake for the unheard of sum of $1,055,000. The nub of the Florence argument was that the firm had staked the fifty-five acres of water-covered ground prior to 1907 but had not been allowed to register them. When governments wish to retain revenue, it is rare that citizens or groups can counter the tax collector. The Florence Mining Company lost its appeal and the verdict lies in a slim pamphlet in the Ontario Legislature. Cobalt was a very litigious camp and civil suit was in process continually.

Cobalt was a tough town with characters to match. Western writer Luke Short visited and saw the annual hand-drilling competition. He summed it up as "the most gut busting sport every devised by man." Ginger beers were not drinks but those who claimed to have mining degrees. Locals felt sorry for such people, who did not seem to know that silver was only found by looking. No other qualifications were deemed necessary. Thomas Edison did not have to prove his genius, but then he did not reside in Cobalt — although he managed an area tungsten mine.

One of the larger-than-life characters was Foghorn Macdonald. His name was apt, as even the slightest whisper uttered by that worthy carried a mile. The most popular story of his exploits relates to shaft sinking. He had a gang sinking a shaft on the outskirts of town. When he left on the occasional trip into town for a "toot" he always measured the depth of the shaft. In this way he could soon calculate downward progress on his return. On one such occasion he went to the bottom, roared for his tape and noted the footage as usual. This time the men made sure a loop was tied in the tape. Foghorn noted a depth of ninety-five feet, then left camp to hit the high spots. On his return, a couple of days later, the calculations were repeated, this time with the tape at its proper length. When Macdonald saw that the tape measured only eighty-one feet, he studied the marker for a while and then stared at the waiting gang. "Now boys," he said, "I know the rock is hard and I don't expect miracles, but I still expect you to hold your own. Get down and make up that lost footage!"

The day shift at noon hour at the Cobalt Townsite Mine. — OA-S 13881-5

As the McKinley-Darragh Mine was being built, the surrounding countryside was bleak.

— OA-S2538

Mine owners loved such pictures. The veins of silver encouraged southern investors.

– Bob Atkinson coln.

The head frame and rock house at the Chambers-Ferland Mine.

– OA-S8610

The Buffalo Mine.   – PA-17807

Machinery from the Beaver Mine was later used at Kirkland Lake, hence Beaver Hill, the name of the incline into the gold town.
                                                                  – Bob Atkinson coln.

50

Everything was gray and all space was utilized. McKinley-Darragh-Savage Mine, 1910.
— Rev. R.W. Lawrence, OA-9160-S13657

The Temiskaming Mine and mill.
— PA-17812

A well run outfit, the Nova Scotia Mine, 1909.
– OA-15380-30

In Cobalt, appearances were ignored in favour of silver production. The Trethewey Mine.

– OA-14898-10

The Cobalt Lake Mill took on a new look in winter.   – PA-17814

The Foster vein was famous but most surface showings in Cobalt were carefully stripped to see where they led.   – Bob Atkinson coln.

Lord Beresford's visit to examine native silver was just one in a stream of dignitaries. Note the water bucket. When wet, silver veins were easier to see.
       – The Standard, Montreal, September 25, 1909. PAC C-22781

Even in Cobalt where spectacular finds of silver were common, this chunk merited attention in 1906.
                                                                                    – PAC-30215

Drilling in a drift. Note the absence of any safety equipment.
– OA-S 13751

The Larose Consolidated Mine was one of Cobalt's more famous properties.
– OA-S8601

Visitors by rail in 1907 found the snow brightened up the usual Cobalt grey. — OA-1450-S926

Jake Englehart, Chairman of the government railway, was quick to recognize the needs of mining camps. — Imperial Oil

Cobalt could boast both electric and steam service in 1910. — OA-S12707

In this view, men were aware that history was being made as the first automobile arrived in Cobalt. – OA-S12439

Cobalt was rugged but offered all professional services, July 1907. – Bob Atkinson coln.

In June 1909, while Cobalt streets appeared prosperous, the bush was close at hand.
— W.R. Forder

This one side of Cobalt's 'Square' used all available space for advertising, September 1909.
— Bob Atkinson coln.

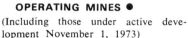

OPERATING MINES ●
(Including those under active development November 1, 1973)

Agnico-Eagle Mines Limited
1. Trout Lake (outside map area)
2. Canadian Keeley (outside map area)
3. Penn Mill
4. Temiskaming

Teck Corporation Limited
5. Silverfields Division

Canadaka Mines Limited
6. Bailey Mill
7. Giroux Lake
8. Conisil

NON-OPERATING MINES ○
(Including dormant properties, past
producers, and properties awaiting
development)

9. LaRose Mill
10. Nipissing 96 Shaft
11. Penn-Canadian
12. Christopher
13. Cobalt Lode
14. Beaver
15. Buffalo
16. O'Brien
17. Colonial
18. Crown Reserve
19. McKinley-Darragh
20. Nipissing
21. Cart Lake
22. Right-of-Way
23. Nipissing 407
24. Townsite
25. Violet
26. Kerr Lake
27. Lawson
28. Louada (Nu-Silco)
29. Patricia Silver
30. King Edward
31. Trinova (Silver Town)
32. Brady Lake
33. Chambers-Fernland
34. Hargrave
35. LaRose
36. Princess
37. Coleman Township Property
38. Coniagas
39. Drummond
40. Hudson Bay
41. Silver Regent - Genesee
42. Silver Monarch

43. Deer Horn
44. Silver Leaf
45. Mentor
46. Pittsonto Mining Co. Ltd.
47. Professor Silver
48. Rockzone (Smith-Cobalt)
49. Silver Cross
50. Silver Cliff
51. United Cobalt
52. Silver Pack
53. Sudbury Contact

54. Chitaroni Minerals
55. Silver Summit

Outside Map Area
Agaunico
Bomont
Kirkland Townsite
Larum
Millerfields
Langis
Louanna

Mines of the Cobalt Camp.

– Ontario Ministry of Natural Resources

O.D.M. 12046

Map labels: NORTH COBALT, Cobalt sediments, Graham L., Sasaginaga Lake, COBALT, Cobalt L., Peterson, Keewatin, O.N.R., Bucke Tp., Fralecks Pond, Coleman Tp., Lorrain Tp., Crosswise Lake, Nipissing, Kirk Cr., Kirk Lake, diabase, Giroux L., Giroux Cr., Coleman Tp., Gillies Limit, Keewatin, New Lake, Lorrain granite

60

Peak and Slow Down

It behoves a man to know the company he is in.
— Jake Englehart speaking in 1911 about wildcat
promotion schemes in Cobalt.

With falling cadence
The muffled bands will strike up
God d _ _ _ the Minister of Mines.
— James H. Tighe of Larder City deplored the
moving of the mining recorder's office from
that town to Matheson.

A street map of 1910 offered something for all in Cobalt. There were theatres and an opera house. Several eateries existed to suit all pocketbooks, including the Uwanta Lunch. All services were provided, and while "blind pigs" were not shown by name, a note on the map assured the thirsty that there were nine in the Hunter Block alone. With churches, clubs and fraternities, the town had all the opportunities of larger communities. The town also paid the price of success. Building lots were a staggering $4,000 each. One added convenience was the newly opened Nipissing Central Railway. A source of competition until the steam railway took it over, the electric line offered frequent transport over the short haul from Cobalt to New Liskeard. The line was the most northerly of its kind in Canada.

Both town and mines now rejoiced in cheaper power. Ragged Chutes, Hound Chutes, Fountain Falls and the Mattabechewan all added power to the grid. The Cobalt Hydraulic Power Company, eighteen miles from town, even offered a scenic attraction with its compressed air blow off. Power costs were reduced from $150-$175 per horsepower a year down to an affordable $50. The new price helped both marginal mines and big producers, as lower ore values could be developed.

Cobalt was approaching its peak period as a camp in 1910, when silver was only just over sixteen cents an ounce. Despite attempts to streamline recovery work, it is estimated that from two to six ounces of the precious metal were lost in the processing over every ton of ore. As the area settled in to the idea of lower returns from area properties, the prospectors just looked harder. Mines like the Big Six or Le Heap out on the Montreal River were losers, but others did well. In town the concentrators kept up their steady roar, processing 665 tons a day. There was more consolidation, with the Mining Corporation of Canada acquiring the City of Cobalt, Townsite, Townsite Extension and later the Buffalo mines. At the same time, the Timmins interests were pulling out. They sold the Larose Mine for $1 million and used the funds to finance a new property to the north. In seven years the Larose had made $8 million and the merchants from Mattawa had reaped a good return on their initial investment.

All the riches of Cobalt came above the 1,600-foot level, and at their peaks most mines were at less than half that depth. Mines like the Kerr Lake were making $10,000 in profit a month. The great Nipissing set the pace. It had ten shafts mining ore from three vein systems, the number sixty-four (or Kendall), the Meyer and the Fourth of July. The latter had been found back in 1907 when a squatter had been moved from a shack which sat right on top of the vein. The "Nip" made $2,736,132 in 1910 and paid twenty-seven

61

The Nipissing Mine mill overlooking the unattractive residue dump, 1918. — PA-13670

M.J. O'Brien was at the height of his fortunes in 1917. — PA-2454

per cent of its capital in dividends. It used hydraulic equipment to strip overburden and later began the construction of an overhead tramway across the lake. A steel net was thoughtfully provided to protect passersby walking beneath the buckets. The big mine used its low-grade mill to process ground often ignored by other mines.

Cobalt never mined out its people. George Caldbick was promoted to Inspector of the Northern Division of the Ontario Provincial Police. M.J. O'Brien bought out his original partner and could reflect on a profit of $2,800,000 in just four years from the mine that bore his name. O'Brien was interested in many projects, and he said later his Cobalt property was " . . . just something to fall back on" if times were ever hard. The men who had made their money in the Larose Mine stare out contentedly from a photograph of the time. Louis Timmins is elegant in a fur coat, while brother Noah fixes the camera as he did men, with a stern, steady gaze. John McMartin had a heavy mustache and looked almost a pugilist compared to the softer features of his brother Duncan. Local mine-made millionaires included Angus McKelvie and Thomas McCamus. Their mine brought in more than $6 million and they went on to success in lumber and telephone services; their names are on streets in several towns.

There were some sour notes. In this time of prosperity Cobalt averaged twelve fatalities per thousand workers. Miners still worked nine or ten hours a day. They laboured in a camp where it was not uncommon for mines to protect their interests with detectives. A shift boss could be fired for criticizing management and any hint of union activity was always of interest to owners. Miners could be as affected by speculation as anyone else and even managers would buy shares in their own mines to promote trading values. Perhaps the most extreme case was that of promoters in Atlanta, Georgia, who wrote a book for suckers about King Solomon's Mines and then tied Cobalt in with it. The author never did see the silver centre and was jailed for his pains.

Cobalt was a strong producer from 1907 to 1920, but excelled in 1911. Even at fifty-three cents an ounce, the 30 million ounces of the precious metal in a year made for an impressive recovery. Noisy, dirty, grey little Cobalt proved the old north-of-England saying: "Where there's muck there's brass." There was also fire. Only two years after the 1909 fire, several business properties were destroyed in the commercial section of town. With the inevitable interest in the new gold camp at Porcupine and closure of marginal mines, the progress of rebuilding slowed in Cobalt.

One very bright spot was the merger of the Cobalt and Hydraulic power companies to form the Northern Ontario Light and Power Company. This new, strengthened firm, with the Mines Power Company, both on the Montreal River, continued to offer greater economies to cost-conscious properties. The Ragged Chutes plant, with its twenty-one miles of pipes snaking around the camp and weekly air blowoff, was a wonder in its own right.

Thirty-five hundred men worked in twelve dividend-paying mines in 1912. The town itself had a population anywhere between 10,000 and 15,000, depending on which source is believed. The Grand Trunk Railway made statements which were hard to swallow. "Cobalt is a staid and settled institution," it declared in a pamphlet of the time. At the same period the T. & N.O. had a demonstration car travelling in Southern Ontario in an attempt to fulfill the line's original mandate to lure settlers north. The railway had little luck, at the time, convincing newcomers that agriculture could be more profitable than mining.

Major Eddie Holland is another man worthy of note. He was an early settler in Haileybury before the advent of the railway linked C.C. Farr's small colony with the rest of the province. The "major" was legitimate, for Holland had won a Victorian Cross during the Boer War. He later became Cobalt's postmaster, but revealed himself as a true northerner when the Gillies Limit finally came open for staking. The whole area south of town no longer had Provincial Mines stamped across it on maps. Holland and his friends knew that there would be a stampede to stake, even though many of the locals were secretly convinced that the provincial ground was likely to be a fizzle as far as precious metals were concerned. The enterprising ex-soldier took dummy dynamite sticks made of sawdust with real fuses and used them to scatter crowds when he went to try his luck in the new territory. The same dodge helped him register the claims in Haileybury. The ploy did not net him any bargain, for the ground lacked silver, but his ingenuity earned Eddie Holland a place in prospecting lore.

At this time some Cobalters lost out on a great opportunity. There was a new camp to the north, at a place called Bell's Siding, which we know today as Swastika. Around Kirkland Lake, to the east, there were some interesting gold properties. Charlie Dennison went north on the train to check out first the Tough-Oakes property and later the Lakeshore. He was not impressed over several visits, and yet the latter became the biggest gold mine in the western hemisphere. The Nipissing Mine also sent people up to look at the Teck Hughes property. Some samples from those claims were shipped back to the silver town in an uninsured crate and the option to purchase was never taken up. The lack of interest merely illustrates the ever present factor of luck in mining. Meanwhile the man who started the Cobalt rush was visited in his Queen's Park office by a seedy-looking character who was distinguished only by the dollar bill pinned to his lapel. Willet Miller ignored the visitor, who eventually left without saying a word. The provincial geologist never did find out the meaning of the visit, but if it had been intended as some form of signal for bribery, the man was very definitely a poor judge of character.

Dr. Willet Green Miller, Ontario's
first provincial geologist.
He named Cobalt.
– Ontario Ministry of Natural Resources

The provincial railway made $80,000 in 1913 from royalties on ore mined from under its roadbed. Area miners received their long-sought nine-hour days, but the main topic for the year was the draining of Cobalt Lake. Several mines had an interest in the lake, the most prominent being the Cobalt Lake Mine, and to a lesser extent the McKinley-Darragh. There is a faded pamphlet in the Ontario Legislative Library today which was written by a Cobalt doctor extolling the virtues of removing the water from the lake. He refuted local citizens who thought that putting the pumps to work would result in a fever epidemic. Were there not fish in the lake and did not gulls eat them without ill effect, he argued. The good doctor dismissed those who whined darkly that the lake was used as a clandestine cemetery. He reminded readers that the waters were a gift of nature, but since they were now polluted there was no real drawback to draining the lake. Finally, he asserted, the old lake bed could be used for recreational purposes.

By 1914 the Cobalt camp had produced five times the construction costs of the provincial railway from North Bay to Cochrane. Silver was always up and down. Prices dropped from fifty-eight cents in 1913 to half a dollar, and then forty-six cents the following year. This just spurred mine managers to greater production efforts and 16 million ounces were mined to counter lower prices. At this point the Nipissing Mine was not suffering, as it only required nineteen cents to produce one ounce of silver.

The first year of World War I saw a considerable drain of Cobalters off to fight for king and country. The new eight hours a day for the industry would have to wait four years before it was of benefit to many miners. Skilled labour became scarce, commodity prices rose, and mining took a back seat for the first time since the start of the camp. The population started to dwindle, but Cobalt still captured the headlines. Its miners used their tunnelling skills on the Western Front.

The total value of silver shipped from Cobalt in the eleven years since it was first discovered reached a staggering $122,750,000. The lake was finally pumped out. Pumps ran night and day forcing water through a 3,600-foot pipeline into Farr Creek, and from there to Lake Temiskaming. The new newspaper, *The Northern Miner*, reported on these and other events. Based at first in the silver town, the paper would become Canada's premier mining news source.

In 1916 silver jumped to seventy-seven cents an ounce, so even marginal mines that had men stayed open. The Nipissing Mine had paid $15 million in dividends so far. The new Workmen's Compensation Act passed almost unnoticed as managers offered a quarter an hour raise to miners, plus a bonus of a similar sum per day if silver rose a certain amount. Not all the mine owners were interested solely in making money. M.J. O'Brien contacted War Minister Sir Sam Steele and used his Cobalt gains to finance an army battalion involved in one of O'Brien's main interests, railway construction. The gift was a very practical contribution to the war effort.

When the war ended Cobalt had passed another milestone. Ten thousand tons of pure silver had come from the camp to date. But the population after the troops returned was down to 7,000 or 8,000. Peace celebrations were dampened a bit, as the legal limit for alcoholic beverages was only a rather watery two per cent. Good stuff at a respectable three times that amount often passed thirsty lips, but government inspectors were ever vigilant. In one seizure many barrels of good suds were dumped in the Montreal River.

Cobalt welcomed returned soldiers in 1919 with a triumphal arch of rough lumber and spruce boughs.
– PA-57676

The first postwar year saw a very special visitor to Cobalt. Edward the Prince of Wales toured Canada, and the silver town was a natural highlight of his excursion. A journalist travelling with him was both poetic and realistic. "Cobalt," he wrote, "is a fantasy town. It is a Rackham drawing with all its little grey houses perched on queer shelves and masses of greeny grey rock. The streets are whimsical. They wander up and down levels and in and out of houses, and sometimes they are roads and sometimes they are stairs." He was also prepared to show warts and all. Cobalt was surrounded by " . . . greeny grey slimes, mill refuse, mills and corrugated sheds. Cobalt is careless how it built itself. Indifference to planning is what makes the place!"

Prince Edward was welcomed by street banners like GLAD U CAME and THE TOWN IS YOURS: PAINT IT RED OR ANY OLD COLOUR YOU LIKE. A dutiful tourist, the royal visitor was shown through the Coniagas mill, where he was given an appropriate memento by two little girls, a thirty-five-ounce bar of silver. At the O'Brien Mine he donned overshoes, oilskin coat and sou'wester hat and visited the three-hundred-foot level. The miners cheered him, and the future king went to a newly opened stope named in his honour. The *Cobalt Daily Nugget* reported that he met "Dick the Nigger" while he was underground and Dick let it be known that he was a British citizen and, in his own words, "no furriner."

There was a side to Cobalt that Edward did not see. The prince was gone and the triumphal arch welcoming returned veterans taken down when the silver town joined in the general wave of labour unrest sweeping Canada. The Mine Mill and Smelter Workers Union represented the 2,500 mine employees of the area. They sought better wages, conditions and a closer say in the town and services offered. The mines controlled medical care, for example, and men thought this should be changed. Although it was not a strike issue, newly returned ex-soldiers feared others would take their jobs. The *Northern Miner* of June 14th raised the spectre of "aliens," Austrians and Germans, taking Canadians' jobs. This was stretched to mean the "red element." In Toronto, *Saturday Night* magazine came out and declared, *Banner of Bolshevism waves over Cobalt."* The strike that year lasted forty-seven days. Production dropped ten per cent over the previous year, but the sanction failed. The companies still kept their central employment bureau, where records of "undesirables" or blacklisted persons were on file. The camp was fading and labour action had come too late.

The great silver centre declined as the Porcupine and Kirkland Lake gold camps came into their own. The Beaver Mine finally shut down in 1920. It had been remarkable in finding 40,000 ounces at the 1,600-foot level, the only really deep spot in the camp. M.J. O'Brien continued to make money and forced the Larose Mine to return ore which had been mined from a drift into the industrialist's property. Cobalt would produce silver for many years to come, but the heady, exciting days were over. In sixteen years the silver metal won from the area's ground had enriched the gross national product by $186,809,746.

The Department of Mines belatedly set up the Temiskaming Testing Laboratories to offer bulk sampling and assay services to the mines in 1922. The facility still operates today, just off the Square. The Larose Mine suffered a major setback when a collapsed section under the railway line flooded the workings. Over the years the mine had acquired so many other properties and these had swelled its coffers. The main vein was 850 feet long, eight to twelve inches wide, and ranged in values from 700 to 14,000 ounces to the ton. Total production to this date was $25 million. The Nipissing Mine continued operation, but after sixteen years paid its last dividend in 1921. Sensing the changing times for the community, the *Daily Nugget* quietly moved south and resurfaced as the *North Bay Daily Nugget.*

The camp mined a respectable $5 million value in 1922, but the big news was the great Haileybury fire. In one day, October 4, 1922, a huge acreage was burned and forty-three died. Haileybury itself was wiped out. Most settlements as far as Englehart to the north and around Cobalt to the south were destroyed. The silver town, victim of so many of its own fires, this time escaped unscathed. Small fires broke out to the west around the Hudson Bay, Coniagas and Trethewey mines, but were easily dampened down. This was the third and last of the great northern fires. The provincial government finally set up the fire service and beefed up all forms of protection. Curiously, the day after the fire, the local paper sported an advertisement for a jeweller's fireproof safe. The offer came too late for those caught in the path of the blaze.

The size of the waste rock dump shows long production at the Crown Reserve Mine. – PA-17815

Cobalt is known as a place where many people made it big. Victor McLaglen is remembered as a man who helped organize boxing matches in town. With his pug nose and heavy features, McLaglen must have sparred in a good number of bouts himself. Cobalters saw the one-time boxer again in 1935 on the silver screen. He played the lead role in John Ford's *The Informer*. This story of the Irish Rebellion was good enough, but Victor McLaglen as Gypo Nolan, the outcast who betrayed his best friend to the Black and Tans for a twenty pounds reward, was superb. He gained a coveted Oscar for the role.

Follow some more Cobalt people. Prospectors still used the silver town and its neighbours to the north as a base. Ed Horne had ranged the bush and been disappointed many times. He even had a stake in Kirkland Lake, but it was a worthless plot sandwiched between two great producers. Finally he made his mark in 1923 across the great lake in Quebec, at a place which would become known as Noranda. Alex Mosher never made much from his Cobalt properties but felt his mining licence B14444 was lucky, as it recalled four of a kind in poker. His Cobalt-learned skills took him all over Ontario and Quebec, and he finally hit the jackpot in Central Patricia, Consolidated Mosher and others.

The T. & N.O. put a spur line into Silver Centre in 1924 and that camp enjoyed a rail link for seven years before the silver petered out and the tracks were torn up. As more and more mines closed, properties were often leased to local miners on a percentage of ore recovered basis. It was a good idea and kept the area alive for years while other prospects developed.

David Dunlap died in 1924. The lawyer maintained his connections with the Timmins family right through from Cobalt to the Porcupine, and his loyalty paid off in handsome rewards. In turn, he left a quarter of a million dollars to Victoria College and $100,000 to the University of Toronto.

The South Lorrain area produced silver from 1908 to 1932, with 1926 the peak year at the Keeley Mine, declaring 1,705,531 ounces of silver recovered, and 1,104,597 ounces at the Frontier Mine in the same period. Cobalt remained the champion, giving an average of 3 million to 5 million ounces of silver delivered a year from 1909 to 1926.

The mine was surrounded by bush. Frontier Mine, South Lorrain. — PAC, P-13643

The close of the decade saw more progress in mining as a failing camp passed on the torch to new properties. The government railway ran its first train from Swastika through to the new gold fields at Noranda (discovered by Ed Horne and friends). The great Larose Mine finally called it quits in 1930, after having given up 25 million ounces of silver and paying more than $8 million in dividends. The Nipissing reached the end of its silver lode the same year but still boasted a $3,500,000 treasury. Thayer Lindsay, the great mining owner-promoter, bought into the company and persuaded the directors to invest the funds in a medium- to low-grade gold property in Northwestern Quebec. The Beattie Mine would produce for over thirty years. Once again Cobalt had aided the start-up of another mine.

Over the years, from the thirties to the present, Cobalt experienced the typical mixture of ups and downs that is the lot of any mining town.

Seven years after his death, David Dunlap's family followed him in philanthropy and gave the observatory which bears his name in Toronto.

Most mines rolled over and played dead in 1932, with only one in operation. To this date $280 million of silver had been recovered and lease arrangements were still going on. The Cobalt Lake Mine had a death blow when it broke through the old lake floor and 400,000 tons of slimes poured through the workings. Total output had been about 5 million ounces. The initial large sum spent to secure lake bed rights had been handsomely repaid.

A diver going down to inspect mine coffer dams on Cobalt lake.
– OA-S2540

At a time when silver prices were once more low, a distinguished visitor came to Cobalt. Governor-General Lord Bessborough came to pay homage to the memory of William Henry Drummond. The poet's house was gone, but the chimney remained and a plaque was inserted in the stones honouring the man who wrote of the habitant and gave his life serving his fellow man in Cobalt.

By 1936 the rich area deposits had been skimmed off and it was obvious all future recovery of silver and other host minerals would be expensive. Only in stable times could Cobalt make any headway, and the Depression was no time for progress in a faded mining camp. But people there were used to hard times and they weathered the lost years as they did so many other low spots. When the T. & N.O. took up the Nipissing Central's tracks, the electric railway put to rest by the motor car, the event hardly caused a stir in the community.

The Second World War and the period following it revitalized Cobalt. Mill residues were taken up for treatment and surface water returned to the mine-fringed lake. Cancer to that time had been treated with the costly and dangerous radium therapy, but now the formerly unwanted companion to silver, cobalt, was used to treat the dreaded disease in the Cobalt 60 Therapy Unit, more commonly known as the Cobalt Bomb. At last the companion to silver, which had so long been discarded as a nuisance by-product, was coming into its own. Cobalt had always enjoyed some use in enamel as an oxide and as a colouring agent in porcelain and ceramics, but now followed more sophisticated use. Its heat-resistant qualities made it of use in the production of jet engines. In the form of stellite, a cobalt chromium alloy, it was used in the manufacture of cutting tools and was also found superior in some respects to nickel in electroplating.

After the war ten small mines were still shipping both cobalt and silver. The Larose had a reincarnation as Silver Mines Limited in 1949 and until 1957 produced both cobalt and copper. By 1963 (sixtieth anniversary) the camp showed a total of $80 million in dividends for the area to date and nearly 1,185,000 tons of silver and concentrates shipped. Actually the sixties brought a new lease on life for Cobalt, as silver prices rose in 1961 from ninety cents to $1.37 an ounce in just fifteen months. New ore reserves were found and old properties were revitalized with new names. The Silver Summit came from the former Savage property. The Deer Horn in another time had been the Cross Lake O'Brien Mine and the Glen Lake Silver mine grew on the one-time Bailey property. The latter produced 1 million ounces worth of silver in 1962.

Silverfields, a division of Teck Corporation, which itself started in Kirkland Lake, was formed in 1964 to mine and mill abandoned Cobalt mines, and also to locate new reserves. Small outfits, each with a handful of men, worked over abandoned workings and made money out of it. Shafts were sunk to test the theory that silver did not extend below the sill of basement rock. There is no keeping up-to-date in Cobalt. As mines close, new ones spring up with regularity. One needs a lot of optimism in the mining game and it frequently pays off.

Cobalt owes its existence to the railway, which made entry possible and exposed its treasure house for those who took a closer look. The place which was at first called a workingman's camp — as all that was needed for a few years were the simplest of tools with a minimum expenditure — netted an amount so large that, if all the silver and bullion extracted in its first sixty years were placed in boxcars, a one-hundred-mile trainload would be required, stretching all the way from Cobalt to North Bay. The initial high-grade returns found capital for more expensive undertakings and in turn further capital, skilled labour and engineering technology was attracted. Cobalt was the cradle of Canadian hardrock mining and became the initial training ground for the industry. Cobalters fanned out and founded other camps. The nearby Haileybury School of Mines graduated people like Gilbert Labine, Murray Watts and hundreds of technicians for the industry. Today

Cobalt continues to function as a mining town and contributes its people to the mining camps of the world.

Except for a few properties, the noise of the ore cars, the mills and tramways are largely stilled today. But take a look at just a few of Cobalt's contributions to Canada. The camp was the greatest silver mining region in the country and a technology leader for the industry. It was largely responsible for the initial success of the Temiskaming and Northern Ontario Railway, (now the Ontario Northland), established the credibility of the Toronto Stock Exchange and saw the start of the Ontario Provincial Police. Old Cobalt is not done yet. The place is still very much alive and kicking. Come and see.

Prospect Avenue today looks much the same as this view. The building left around the head frame is still there and the hotel still has the suds.
                                                                    – OA-9624

At Cobalt they just followed the vein. Old working of the McKinley-Darragh Mine, 1927.

– PA-13763

The great Carson vein, Crown Reserve Mine, Kerr Lake 1913.   – Ontario Bureau of Mines

An electric locomotive at the main station, 350 foot level, M.J. O'Brian property, Gowganda, 1928.
– PA-17434

A view across the lake to the Cobalt Lake Mine, 1917. The houses were placed as an afterthought.
— OA-13390-43

The head frame was dwarfed by the mill. Penn Canadian Mine, 1918.   — PA-13672

This Cobalt powder magazine of 1924 had more charm than similar structures of concrete.

— PA-86938

The Cobalt Reducing Mill was one of several processing ore.

— Ontario Bureau of Mines, April 1922

Coniagas Mine, Cobalt, Ont.

Coniagas Mine.
– OA-S12430

The Right-of-Way Mine leased silver rights along the railway tracks. – Teck Centennial Library

This view was taken in 1919 from the roof of the Nipissing Mine reducing mill. Note the aerial tramway and the rail station, centre right.   – OA-15808-85

Spectacular nuggets always had to be recorded for posterity.   – Bob Atkinson coln.

This mine, the Casey Cobalt Mine, nine miles from Cobalt, was destroyed by fire August 22, 1915.
– OA-S8612

The Lawson Mine had the famed silver sidewalk.
– OA-14898-30

At Cobalt, the mills dwarfed head frames. The Cobalt Lake Mine. — OA-S5640

The King Edward Mine, formerly the York Ontario. — OA-14898-15

The Cobalt Flier. There were several fires and the one referred to on the caption was likely 1912.
– Cobalt Museum

Cobalt was the southern terminus of the Nipissing Central Railway. Miners found the service
most convenient.
– OA-14898-19

SIR WILFRED LAURIER AT COBALT. SEP. 18th/12. PHOTO MACLEAN.

The federal opposition leader passes Cobalt station.
— PAC, MacLean, C-37307

Cobalt station was the key social spot in town.

— Author coln.

Having a hospital sponored by the mines made good business sense. Miners paid a monthly fee for service.
– OA-1489-25

Some wag named the open intersection above the station as 'The Square'. Like many places in the silver town, it confounds angular measurement.
– OA-15409-6

Cobalt streets climbed hills in 1924. The place is much the same today minus the horses.

– PA-86947

By jumper sleigh and shanks pony men came to look for work in the gold fields. — Author coln.

GOLD IN THE PORCUPINE

Gold, gold, gold, gold!
Bright and yellow, hard and cold;
Molten, graven, hammered, rolled,
Heavy to get, and light to hold;
Stolen, borrowed, squandered, doled.

– Thomas Hood

PROSPECTORS READY
CROSS PORCUPIN
LAKE

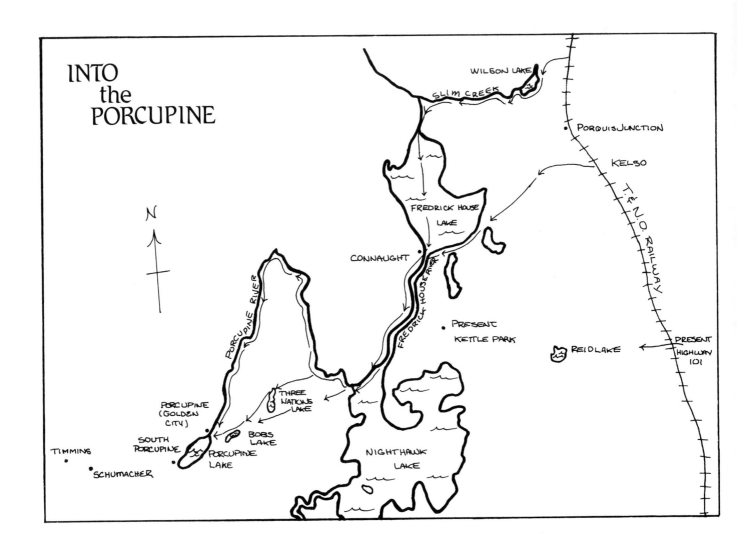

INTO
the
PORCUPINE

WILSON LAKE

SLIM CREEK

PORQUIS JUNCTION

KELSO

FREDRICK HOUSE
LAKE

CONNAUGHT

PORCUPINE RIVER

FREDRICK HOUSE RIVER

PRESENT
KETTLE PARK

REID LAKE

PRESENT
HIGHWAY
101

T. & N.O. RAILWAY

N

THREE
NATIONS
LAKE

PORCUPINE
(GOLDEN
CITY)

SOUTH
PORCUPINE

BOBS
LAKE

PORCUPINE
LAKE

NIGHTHAWK
LAKE

TIMMINS

SCHUMACHER

The Sad Story of Reuben D'Aigle

Quartz veins in Ontario never pay to work.
— Bob Mustard, prospector

One of the great axioms of mining is that the game gives out even-handed chances. The romance lies in the story of lost opportunities. Reuben D'Aigle is the forgotten man of the Porcupine. He had a colossal fortune within his grasp and missed it by a boot print. His name is not well known today in the modern City of Timmins. After all, who recognizes those whom Lady Luck deserts? There are always men like D'Aigle who are ahead of their time, looking for the next strike.

D'Aigle was a young man in the last decade of the nineteenth century. He was from New Brunswick, and like so many others he was lured by the siren call of the Klondike. With other hopefuls, he suffered the six-month journey around South America and Cape Horn. He realized that the Klondike was largely staked out. Reuben D'Aigle differed from his fellow greenhorns in one important way. He was extremely stubborn and rarely gave up without a fight.

He travelled up the Yukon River into Alaska, and while many disillusioned prospectors continued on to the sea and a boat home, D'Aigle went off on his own and prospected the Kayukuk, a tributary of the Yukon solely in Alaska. At Cleary Creek he found a gold camp starting up. River colour looked good and D'Aigle staked thirty claims. One of these became the richest in the camp. When he sold out seven years later, D'Aigle had a big bankroll and so much placer gold he had to trundle it to the waiting sternwheeler in a wheelbarrow.

Guards took the place of the wheelbarrow in Seattle. Once the gold was deposited in the offices of the United States mint, D'Aigle did not follow the usual route of many successful prospectors of the time. He did not lose most of it in a big spending spree. Instead D'Aigle looked around for mining fields. Nothing out west took his interest, and then he heard the news of the silver boom at Cobalt.

It was a case of history repeating itself for D'Aigle. The major ore bodies had been staked out and he saw himself as a prospector rather than a miner. He left the area abruptly and went south to Kingston, Ontario. Queen's University was giving a short course in minerals. D'Aigle felt he needed some formal instruction, as his only experience had been with rivers bearing placer gold, and deposits in the East were much different. While at the university, he spent much time in the library reading all geology and survey reports. He noted the comments of one survey party to the effect that gold-bearing rock had been observed in the Porcupine area. When the course was over, he put his bush gear together and headed north.

The logical place to enter the Porcupine was by the new T. & N.O. Railway to McDougall's Chutes, site of present-day Matheson, and travel overland from some point along the line. But D'Aigle rarely did things the easy way. With a Metis companion, Billy Moore, he took the Canadian Pacific Railway north to Mattagami, a tiny settlement seventy miles west of Sudbury. The whole area is veined with lakes and rivers. D'Aigle and his companion put their canoe in the Spanish River and paddled north to the Mattagami River, where they were able to drift leisurely down the waterway. Then the two portaged and paddled again along a string of lakes between the Mattagami and a large lake called Porcupine. One of these lakes is now Miller Lake. They moved on to Delbert, today Gillies Lake. There was another short portage to Pearl Lake. No sign existed of any particular riches on its shores. They hiked for a one-and-three-quarter-mile portage and another mile of paddling, and they reached the big Porcupine Lake.

As D'Aigle scanned the lake and surveyed the area, the report of the geologist W. Parks of Toronto came to mind. It had been readily available in the university library, but was not exactly a well-thumbed document. This was the area that Parks had indicated held a promise of gold. There was rock everywhere. It lay underfoot covered by moss, jutted out in bare clumps between the trees and presented itself in many forms. From unlined slabs to deeply crevassed outcrops, the ground offered a prospector plenty of work.

Back south of Gillies Lake, a great white knob of quartz jutted up among the trees. D'Aigle and his partner scrambled about it, making easy going in the prospector's hob-nailed boots. There was some visible gold and they staked a total of seven claims. Somehow the gold that D'Aigle saw disappointed him. The quartz seemed to dwarf every other showing. The two men chipped some samples from the claims and went south. The specimens later showed promising results, but did not impress D'Aigle. He was, like so many prospectors, interested only in a spectacular find.

No one every said that Reuben D'Aigle was a quitter. The next summer he used his earnings from the Alaskan claims to finance a larger party into the Porcupine country. This time he took   hand steel and blasting powder. Several small pits were excavated, but there was none of the free gold of which the stubborn prospector dreamed. Bob Mustard, a seasoned prospector who accompanied D'Aigle, summed up their disappointment. "Quartz veins in Ontario never pay to work," he declared. D'Aigle agreed and left the steel and anvil in the pits when he departed.

Reuben D'Aigle spent the rest of his life prospecting, until he was well into his seventies. He was to be the first in another great camp and had the same disappointing results as in the Porcupine. His seven claims lapsed for lack of work and proper registration, and the moss grew up around his test pits. That neglect was too bad because the moss covered a fortune.

Reuben D'Aigle, the 'forgotten' man of the Porcupine in 1959.
– Toronto Star

The Mine Finders

Gold is where you find it.   – Old Saying

Ah, take the cash, and let the credit go.   – Omar Khyayam

The Porcupine story starts before Reuben D'Aigle and his misfortune. North of lakes Temiskaming and Abitibi, a long crescent of mineral-bearing rock stretches several hundred miles. Someone had to stumble across it eventually. Even before the Cobalt strike, E.M. Burwash made a survey for the Department of Mines through the Porcupine country. He found quartz veins bearing gold in Shaw Township, to the south of present-day South Porcupine. Strangely enough, no major finds were made in Shaw. In 1898, 1899 and 1903 a University of Toronto geologist, also working for the provincial government, found gold-bearing rock in Whitney Township. On his rapid survey trip, W. Parks crossed the ground which would become the Paymaster Mine, and was close to the Dome, Broulan, Hallnor and Pamour mine sites. His most oft-quoted comment from 1899 had spurred Reuben D'Aigle: "I regard the region south of the Porcupine trail as giving promise of reward to the prospector." His colleague, Kay, was more pessimistic, saying, "No minerals of economic importance were found." Both men passed Miller Lake, a tiny lake now filled by mine slimes. If they had spent any time there at all, they would have found one of the Hollinger veins. But $400 million worth of gold had to wait a while longer.

Although these gold showings became known among prospectors and investors alike, the money men had doubts about the possibility of a rich Ontario gold field. In the 1880s, the big gold rush to Lake of the Woods had proved disappointing. Others worked their way through the Porcupine in addition to the unlucky Reuben D'Aigle. The three great mines which finally emerged were not the makers of the Porcupine. Others not so fortunate had been there before. E.O. Taylor had camped on Nighthawk Lake three years prior to the major discoveries. That wandering missionary Father Paradis had crisscrossed the area, but his brush with prospecting came later. In 1907 Victor Mattson and Henry Banella had a small mine on Nighthawk Lake. They even rigged up a crude mill, which produced a bar of gold before fire put them out of business. What a lonely life it must have been for men such as these. Every rare visitor was welcome for news and a change of conversation.

It is not necessarily the most successful claims which start a gold rush. George Bannerman and Tom Geddes were prospecting north of Porcupine Lake when they found very good surface showings and started what was later to become the Scottish-Ontario Mine. Bannerman was in the area at the same time as the other prospectors who were to make the rich claims. Like the others he was not prompt in registering his ground in Haileybury because of the distance involved. When he did make the trip south to record his claims, Bannerman was practically mobbed when the crowd saw his samples. The Bannerman-Geddes claims later received backing from investors in Scotland. The original name changed to Canusa (from Canada and U.S.A.), and finally it became the Banner

Porcupine. But for all the name changes, the property did not prove out the early rich surface ore. It petered out underground and the mine was a poor producer. The place was easily forgotten when the rich finds were made.

Jack Wilson was a longtime prospector. When he heard the stories about gold in the Porcupine, he was working for the railway. In 1907, travelling on a handcar with a railway construction foreman near Boston Creek, he stopped the car to view some rock formations and brought out gold samples from the area. The ore looked good, but Wilson could not interest anyone in his samples. In this way he missed being in on the Boston Creek-Kirkland Lake area camps. But his destiny lay further north. He was based with the railway at Driftwood, now Monteith. He contracted typhoid fever late in 1908, but when he recovered he hunted around for backers for a trip into the Porcupine.

Jack Wilson was a great believer in taking a large crew on such trips, as he felt it evened the odds against finding what nature had hidden. He gained backers in two Chicago businessmen, W.S. Edwards and Dr. T.N. Jamieson. Then followed one of the unlucky breaks which so often occurred during the early days of mining in Northern Ontario. Edwards sent $1,000 for the Wilson party at Driftwood. At the prospect of such a deal, Wilson had quit his job and gone back to Cobalt. Driftwood had a blacksmith's ship which also doubled as a post office. The letter fell from the nail where it had been spiked and was only found in the spring melt. Meanwhile Wilson languished in Cobalt until he received word about the missing funds and headed north.

Included in his party were Harry Preston, Cliff and Frank Campbell and George Burns. In mid-May 1909 they took three canoes by train to mile 228 on the T. & N.O. and then worked their way by portage and paddle into the Porcupine, fifteen miles to the west. For the next two weeks it was a steady routine of mapping and prospecting, with good traces of free gold occurring regularly. Then around June 9th the party came across an outcropping or dome of quartz set out prominently amid the surrounding area. The party trenched the dome. Jack Wilson later recalled the turning point in their work: "As I was examining the seams in the quartz, about twelve feet ahead of me I saw a piece of yellow glisten as the sun struck it. It proved to be a very spectacular piece of gold in a thin seam of schist . . . when the boys came back, we got out the drills and hammers, and that night had about 132 pounds of very spectacular specimens!" They had found a vein several hundred feet in length and 150 feet wide.

In later accounts some have it that Harry Preston slipped on the rock and others that he set off a blast of dynamite in the quartz dome. This large vein plastered with gold became known as the Golden Staircase. The gold just hung down the hillside. One visitor later said: "The gold was in blobs like candle drippings, and in sponge-like masses, some of them as large as a cup, lying under the moss . . . The Big Dome, they called it."

The party was elated and Wilson left to wire his backers of their good fortune. He also very quietly went out and obtained new mining licences. Their limit had been exceeded earlier and those four claims which were to be the nucleus of the Dome mines were illegally staked until the new papers arrived. W.S. Edwards was so excited by the original find that he travelled north with Wilson on his return. After an unfortunate start when Edwards, an extremely stout man, upset the canoe in which they were travelling, they reached the camp and the backer was able to feast his eyes upon a veritable jewellery box in the crumbled rock.

A deal was concluded by the light of the campfire. The man from Chicago and the prospectors came to terms. Jamieson and Edwards received fifty per cent for their backing and further assessment work. Wilson took ten per cent, and the other seven in the party shared the forty per cent remaining. There was also an outright cash payment of $1,000 to each man. All agreed except Preston. A bottle of whisky placated him and he went to guard the Golden Staircase against all intruders. Edwards then returned to the railway to

The gold rich quartz is not much to look at on the Dome property but one like this started the gold rush. — OA-S13738

Adits, horizontal tunnels rather than vertical shafts, were easier to mine. Barite mine north east of Timmins. — OA-S15934

inform his partner of the substance of these exciting developments. His telegram reflected the mood of the whole group: "Have discovered the golden pole beyond description stop Answer Matheson Wednesday Haileybury Thursday stop Protect checks six thousand at Massey stop"

Barbershops were great places for prospectors. They heard all the news of the bush and could compare notes. Young Benny Hollinger was a barber and with his friend Alec Gillies had been cutting some pulp wood on the side. When the news of the Bannerman and Wilson finds hit Haileybury, they were in that town trying to rustle up a grubstake for themselves. Backers were hard to find. Gillies managed eventually to get $100 from Jack Miller, while Hollinger only managed $45 from his uncle Jack MacMahon, the bartender at the Matabanick Hotel. So rapid was the speculation in those days, with both the silver boom and now the gold rumours further north, that MacMahon immediately sold half his interest to Gilbert Labine for $55. A rare photograph of Benny Hollinger, taken just before he left, shows a very young man. A bush hat is tipped on the back of his black hair and his gaze is shy and unassuming. It is the face of one lacking in guile, yet from the faded portrait comes a feeling of strength and dependability.

Benny and Alec were bitterly disappointed when they finally packed into the Porcupine and arrived later at the Dome camp only to be told that the ground was all staked. Wilson advised that they go about four miles to the west, where there was open ground. They followed this tip and later found themselves near Pearl Lake. Reuben D'Aigle's rusty anvil and drill steel lay in his small pits where he had abandoned them. Prospectors commonly found evidence that others had visited an area before them. Such things as adits, or horizontal tunnels, trenches, pits and even blasting following a vein were usual relics in the gold country. There was evidence of much foresight and hard work. The rusting anvil and steel gave mute testimony to the effort exerted in packing that weight in, let alone the excavation work. The fact that someone had gone to great trouble and expense in the area made the youthful pair even more careful in their search.

By the toss of a coin Benny staked first and Gillies staked around him. Gillies was on the lower ground and they were out of sight of each other. Gillies later described what happened next. "I was cutting a discovery post, and Benny was pulling moss off the rocks some distance away, when suddenly he let a roar out of him and ran and threw his hat at me. At first I thought he was crazy, but when I came over to where he was, it was not hard to find the reason. The quartz where he had taken off the moss looked as if someone had dripped a candle along it, but instead of wax it was gold. The quartz stood about three feet out of the ground and was about six feet wide with gold splattered all over it for about sixty feet along the vein."

For the next few hours, Gillies and Hollinger staked twelve claims between them. It was a genuine frenzy of activity, for there might be another prospector behind every bush. Being faithful to an old friend, they reserved one of the claims for Barney McEnaney, a former partner now crippled with sciatica. His bush-ranging days were over and the claim his friends gave him ensured that he would never have to work again. Later it became the Porcupine Crown property, before it was consolidated into the Hollinger Mine. Staking done, the tired pair did not celebrate but just rested before the trip south to record the claims. Both remarked on the impressions they had seen in some of the free gold. In one case, a heel print was clearly visible. Poor Reuben D'Aigle! He had tramped all over that very area and maybe even sat where they rested now. Here were his footprints preserved in the gold he had been seeking. Benny and Alec made the long journey out by water, land and rail to Haileybury. A sample taken near the site of one of those sets of boot marks assayed in extremely high values.

When Benny Hollinger and Alec Gillies found gold beyond their wildest dreams in that summer of 1909, others were not far behind. Those fears of prospectors who might stake the ground they had marked out were very real. Clary Dixon, Tom Middleton and Jack Miller packed into the area just as Hollinger and Gillies finished staking. Dixon recalls hearing the excited yells that the two men made in their triumph. As the trio rested in their canoe, the other two prospectors, still flushed with success, joined them. Sandy McIntyre and Hans Buttner had also just pulled in to shore, but they did not stay to learn more about the excitement. As Dixon, Middleton and Miller shoved off to the west, the two latecomers went north a short distance and proceeded to stake the nearest open ground. If a big strike had been made to the south of them, chances were that gold might exist close by.

McIntyre, the Scot, and Buttner, the German, are known mainly for two claims that they staked that evening. Numbers 13307 and 13306 on Pearl Lake would form the nucleus of the third great mine in the Porcupine camp, the McIntyre. Buttner was a stolid immigrant who did not speak English well at the time, but Sandy McIntyre was a very colourful figure and he gained in fame rather than in money, which ran through his fingers as quickly as he made it. There are few good pictures of Sandy left. The most well known shows him dressed as a prospector in the manner of a man for whom the seasons have no barrier. He stands on snowshoes with an axe handle protruding from the pack on his back, a signal of preparation for staking. The thick bush coat is half open and a battered felt hat sits jauntily upon his head. His face is three-quarters covered by a thick beard. Grey and white, it serves to frame the face of a man who has spent years in the bush. The man who was acknowledged by his peers to be a superb bush man stares forward in the picture, ready to hit the road again.

Life was not always so exciting for Sandy. A great talker, he was noticeably reticent about his early life in Scotland. His real name was Alexander Oliphant, one of the family names connected with the Sutherland clan. By trade a moulder or pattern maker, he was an itinerant worker who spent much of his time away from home, working in the industrial northeast of England. It appears that the arrangement was satisfactory as far as his wife was concerned, for their marriage was not a happy one. Oliphant's wife supposedly had a shrill tongue and he found life quieter on the road. He dutifully sent housekeeping money home, but either a lapse on his part or the speed of the royal mails brought Mr. Oliphant to Canada.

One week his wife's share of the pay came late to their home in Glasgow. Mrs. Oliphant sent her husband a blistering letter of complaint, and for the restless Scot it was the last straw. He abruptly took passage for Canada, adopted his mother's maiden name and started out fresh as Sandy McIntyre. For a while he worked for the T. & N.O., but soon left in favour of the prospector's free life. He passed Cobalt by and in 1906 squatted at what is now Bourkes, south of Ramore, and gave the place the name Scotty Spring. But Sandy was always on the move. He prospected in Larder Lake and his licence K51 came from that mining division. Since Hans Buttner's licence was 7283B from Haileybury, their association was likely a new one when they found gold in the Porcupine. Buttner disappeared after selling his share. Sandy sold first a quarter of his interest for $300, then an eighth for $25, and later a half for $5,000 and an option for $60,000, which he never collected. From that time forward, Sandy's fortunes continually fluctuated. He was never really lucky again.

In later years Clary Dixon recalled that day in June when he and Tom Middleton had come across Benny Hollinger yelling at the top of his voice, the pressure of staking over and realization of what he had found just beginning to dawn on him. Alec Gillies showed him a piece of rock he had chipped from the hillside. It was about three and a half by two and a half inches and appeared to be seventy-five per cent pure gold.

Sandy McIntyre and his partner had gone north, while Dixon and Middleton saw free gold everywhere as they staked to the west. Their six claims were later purchased by thc Hollinger interests, and related mines were the Middleton and Coniaurum. Clary Dixon was only eighteen when he co-staked the property. Jack Miller, who had thrown in a canoe and supplies to account for $100 worth of grubstaking, became a millionaire from the subsequent sale. By October the main ground was staked. The news of the Hollinger, Dome and McIntyre discoveries had leaked to the outside world and the rush to the Porcupine was on.

These men took a hard route to find their personal eldorados.
– Author coln.

The lead man stopped to take the picture of the Porcupine-bound hopefuls.   – Author coln.

This packer into the Porcupine had a tumpline from his pack to his forehead to support the extra weight. — OA-3843

It was not hard to see why a rail connection was eagerly sought. — Author coln.

The sign was fancy but the big companies did their own assay work or shipped it out; that way secrets were kept. — Bob Atkinson coln.

Porcupine character Johnny Jones once took a dog sleigh down Toronto's Yonge Street loaded with a present for the mayor of Hudson's Bay coal. — Timmins Museum

Her name is unknown but she dressed for the weather. — Bob Atkinson coln.

The mood may seem casual but the supply packs mean business.   – H. Peters, Bob Atkinson coln.

– Bob Atkinson coln.

Alex Gillies and Benny Hollinger at Timmins, 1910.   – OA-S3068-17

Development and Disaster

Men have a touchstone whereby to try gold, but gold is the touchstone whereby to try men.
— Thomas Fuller, *Holy State*

The birds never brought all this gold here.
— John Hays Drummond

Some years ago a heavy flat table of rock was found in the bush near Timmins. There was a hole in the middle and scattered around were some large chunks of rock. There was all sorts of speculation as to its use, but the answer finally came from a student of the Middle East. The set up was an Arrastre. Some long-forgotten exploration people had copied an idea from ancient Egypt. Heavy rocks were dragged over the table stone until the material was fine enough to go to a mill for processing. Mining has always followed a similar route, making use of good ideas wherever they may be found. The geologists who followed the prospectors discovered that the Porcupine area was much altered and metamorphosed. The camp consisted of great flows of lava, dacite, andesite and, rarely, basalt. Cyril Knight made the first map of the Hollinger property and at that time found it difficult to draw boundaries between the intrusive porphyry and the basic lavas. As he went about his work, he noticed D'Aigle's original pit in the heart of the Hollinger veins, with the abandoned forge a mute testimony to one man's failure.

With the staking of the three great properties, the Porcupine came alive as hundreds of canoes bearing prospectors and just plain green adventure-seekers descended upon the area. Golden City and Pottsville sprung up, with South Porcupine soon to follow. Many properties were staked and several became producing mines within the next five years. Such mines as the Mace, Vipond, Coniaurum, Anglo-Huronian, Northcrown, Carium, Newray, West Dome and Porcupine Crown were just a few that were early properties. Some later amalgamated. The early spate of producing mines was not a flash in the pan. The Porcupine area has seen many new mines, and there has never been a period since 1911 when there were not gold mines operating in the area.

The gold-bearing area was later found to be three miles wide and five miles long. Newcomers like Eddie Holland and Jack Leckie were so captivated by the exuberance of the place that they wrote a song about it. They had landed in a camp which produced $35,559 worth of gold in 1910 and would never ship so low a value again. It is worthwhile examining just how difficult it was for people to reach the camp then, and yet many came despite the hardship. Take Noah Timmins. When he was involved in sizing up the camp, he left from mile 222 on the T. & N.O. with a total party of twenty-eight men. It was an unusually cold fall. They made up camps and started cutting a road to Nighthawk Lake. After a while it was discovered that a better way lay through some old roads left by lumber operators. Though it was the longest way round, it was the shortest and easiest route to their own destination. The next morning they used the two teams of horses, hauling freight for the party, to plough a way through three miles of snow to the big lake. The lake had

99

Freighting into the Porcupine.   – Tomkinson

frozen early, but there were only two inches of ice and a total of twelve miles to cross. The load had to be rearranged and the teamsters were not happy with the prospect, but by dint of several trips the passage was completed. Horses still broke through the ice at least a dozen times. Next came a twenty-mile passage to Pearl Lake, but a rough road was opened up by New Year's Day 1910.

Noah Timmins was involved in that operation every inch of the way, and his hard work and effort clearly illustrates the reason why he was already a wealthy man. The drive and determination he had displayed at the opening of the Cobalt camp now ensured his dominance of a second great camp. Contrast the Timmins experience with that of another traveller just a few months later. K.P. Bernhard, a mining engineer, arrived at Kelso, where an old day coach served as a station house. "At Kelso one had to stay overnight in a stopping place. They were too modest to call these places hotels. The walls between the rooms were of rough lumber and one could see through the cracks into adjoining rooms. The only conveniences in the rooms were a water pitcher, a basin and a 'john'." Bernhard went on to say that he continued the expedition via corduroy road, launch, two canoe trips and two portages, which themselves came to ten miles. There had to be an easier way to the gold fields.

Despite the very obvious riches of the Porcupine, financing was not easily arranged in every case. W.S. Edwards set up his headquarters at the King Edward Hotel in Toronto but spent a long time waiting for investors to beat a path to his door. Several people heard his presentation and a few went to see the property. One was Captain Joseph Delamar, or de la Mar, who had come north before to inspect the Cobalt camp. Now he represented the Morell Syndicate, which was heavily connected with the International Nickel group in Sudbury. One account has it that the captain looked over the Dome and strongly recommended it to his main principal, Ambrose Monell. The latter entered Porcupine lore when he said, "We'll take it!" — which was considered a very promising statement, for the financier very rarely made such lengthy speeches.

VEIN OF GOLD 4 FT WIDE
ON BIG DOME

Father Paradis and Corneluis Hurley on a four foot wide gold vein, Dome Mine, 1910.
– Timmins Museum

The Dome property was first worked as an open pit. The rich surface showings were quarried with comparative ease and an armed guard was posted to protect the exposed ore. Many said that the knob outcropping was a freak, but as the mine went underground its value was proved. Harry Preston became excited at the doubt placed on the discovery, but a famous American mining engineer reassured him. "The birds never brought all this gold here," said John Hays Drummond. Drummond knew his precious metals, for he had made enough money in Cobalt to keep interested. Jack Wilson, Edwards and Dr. Jamieson became chief officers in the new company, but it was controlled by American interests from the start. As well as Delamar and Monell, "Handsome" Charlie Dennison of Cobalt (and later Kirkland Lake) fame was also one of the principals. De la Mar was their representative in the camp. An adventurer who, like the Timmins brothers, had initially made his money in silver, he earned the nickname "Hard Cash," for his usual method of payment, no matter how large the amount.

If the glaciers had eroded another 150 feet, there would have been no Dome outcropping. The Golden Staircase disappeared at the 200-foot level and the Dome settled down after that for regular hardrock mining. The company capitalized in 1910 at $2.5 million, based on shares at a respectable $10 each. The first year 247 tons of high-grade ore were produced and made it obvious the Dome would be around for a long time.

The McIntyre, the third largest of the great mines, never had the rich surface showings of either the Hollinger or the Dome. Hans Buttner left at once, but Sandy McIntyre hung around for a while until he lost most of what he had made on his great discovery. The McIntyre had a chequered history before it became established as a mine. Charles Flynn had it first, then A. Freeman, and later J.P. Bickell. Actually only two of Sandy's original claims were incorporated into the mine. It was under Bickell that the ground was put together which would stand as the main property. It was not until 1912 that the McIntyre Mine was incorporated. It had a share capital of $3 million, all of which first sold for a dollar a share. By then the man who found the property was long gone and was to rue the day he got out of his big find so cheaply.

Alphonse Paré was the nephew of Noah Timmins. The young mining engineer had heard of the Porcupine camp and it was he who informed his uncle initially of the great strikes there. He made a fast trip to the Porcupine, and beyond making a confirmation of the vast potential of the area, he did not make any deals but returned to report to the family. We have seen how Noah Timmins responded and made a trip in force into the camp to see the place for himself. That excursion was not based just on a speculative venture. He was already committed with ex-bartender John MacMahon, Hollinger's original grubstaker, for $2,000 down plus larger payments every sixty days until a deal was made. In the meantime Jack Miller had optioned his claims in the Gillies property, now called the Acme, to the Timmins' arch rival of Cobalt days, M.J. O'Brien. The deal was for $50,000 to be paid in two months and a balance of $200,000 on closure. It was just the sort of arrangement the wily O'Brien enjoyed. He instructed his engineer, a man named Culbertson to do some drilling quickly before the initial payment was due. Culbertson did so and then packed out with his samples. The load was heavy and so he selected only those that had quartz in them. Assay results were poor and O'Brien dropped his option when Miller refused a time extension. That was too bad because it turned out later that the stuff Culbertson left behind carried the real values.

Noah Timmins was still up in the Porcupine after his hectic trip in to the Hollinger ground. He worked alongside the men building cabins and then stripping the number one vein and initial shaft work. There were frequent visitors, one of whom was an engineer named H.N. Plate, who grandly informed all and sundry that the Hollinger would never return the money spent on its exploration. That worthy never figured very much in the Porcupine after that. Noah Timmins had heard that Monell had offered Miller $400,000 for his property, but it was to come in relatively small installments. He countered that with $350,000, one seventh of which was to be paid on signing and the balance in payments due every sixty days. The McMartin brothers and David Dunlap once more joined him as partners in this venture.

En route to Toronto to complete the deal which Wilson finally accepted, Noah met Alec Gillies on the train, and the prospector volunteered, "O'Brien overplayed his hand on the deal and lost out. But there's another son of a gun has agreed to put up fifty thousand dollars first payment for a better deal." The identity of the "son of a gun" had been kept secret and Timmins chuckled to himself at this grudging testimonial to his business acumen. When he arrived at Haileybury he ran into Hugh Sutherland, a Toronto promoter, who greeted him with "I have two or three claims not far from the Dome. Would you like to buy them? I'll take $5,000." "Sold!" said Timmins. "I'll take them." Within a short time the Timmins interests had acquired 560 acres made up of the Hollinger, Miller and Gillies four claims, the four Millerton claims and the three Acme Gold claims. The total purchase price was a record sum for gold properties at the time. One million dollars secured a mining empire. The Hollinger was incorporated at $3 million, comprising 600,000 shares at $5 each. Little had Benny Hollinger dreamed six months before when he put down a tiny bucket on a windlass in a test hole that his claims would be the nucleus of so large a mine.

The Timmins group formed the Canadian Mining and Finance Company Limited to represent their syndicate. There were several claims the syndicate acquired that were not within the scope of the Hollinger and these were later sold to Berwick Moreing of England. This company exploited its new ground but missed out on a fortune. The McIntyre was on the block at that time for a mere $90,000 and Timmins suggested that the two companies get together. If the British firm purchased the McIntyre, the Hollinger could throw in two adjacent claims to make a viable mine and work the property on a fifty-fifty basis. Noah Timmins said later, "Mr. Moreing rejected this suggestion, and the magnitude of his mistake may be gauged from the fact that the McIntyre has paid (up to

PEARL LAKE   SCHUMACHER TOWN-SITE   BUNK-HOUSE   CLUB-HOUSE   McINTYRE GOLD MINE   SHAFT   STAMP-MILL

H.PETERS PHOTO

Of this view, only Pearl Lake is unchanged today.   – Bob Atkinson coln.

1936) twenty million dollars in dividends, while our addition to this would have been double that."

Development of the camp was hampered by transport costs. Right up to the late spring of 1911, a stage ran from Kelso to Fredrickhouse and Crawford's Landing and then a rough road into the Porcupine. Freight expenses were reflected in higher wages. Surface workers received $2.50 a day, while underground miners took home $3. By May 1910 there were 750 men working in the main camp and the province recognized this by the end of the year with a road from Hill's Landing to Golden City. As the new year opened, there were up to fifty stages operating into the gold camp and all were crowded.

T. & N.O. chairman Jake Englehart ordered a full-scale survey of a probable rail link into the Porcupine in the spring of 1910. The mining engineer retained by the provincial railway firmly supported the need for a line to connect with the main Cochrane to North Bay track. He said, " . . . the properties are held by experienced mining men from whom could be expected a progressive, and at the same time, rational development." The eventual line was just over thirty miles. The junction came just two miles north of Kelso. The new Abitibi Pulp and Paper Company at nearby Iroquois Falls contributed part of the name. Combining Porcupine with Iroquois provided the name Porquis Junction. The price tag was $450,000 and the line was complete in six months. Construction was tough during the winter months and manpower short, with the old story of precious metals luring railworkers to seemingly better prospects. The province countered this trend by bringing up prisoners serving light sentences in provincial jails to work on snow removal and other chores which detracted from the forward building purpose of the railway construction crews.

The iron road reached the Fredrickhouse River in April 1911, and by June 7 the steel arrived at Golden City, known simply as Porcupine. The official opening took place on July 1st, Canada's birthday, despite a still rough-and-ready track bed. Invited guests stared at the camera soberly in a photograph of the event. Emotion was not considered a pictorial subject. Chairman Englehart toasted the railway and its new association with the Porcupine in a reception following the arrival of the first official train. Neither he nor his audience had any notion of another role that the railway would play in just a few short weeks. There were 8,000 claims in good standing in the camp that summer and the place was a hive of activity, but soon practically all man's efforts would be wiped out in less than a day.

Disaster in mining camps usually means explosions, floods or cave-ins below surface. Occasionally a big underground fire, such as the camp was to experience twenty years later, will retard development. In this case the new camp was to be caught up in a natural holocaust which ravaged a huge chunk of Northern Ontario. Snowfall that year had been light and there was little water in the bush. This was promising for local development and the roads were passable sooner. The bush was full of prospectors and small fires flared up, some due to carelessly covered campfires. In mid-May 1911 the new Hollinger surface plant was destroyed by fire. The pick and shovel brigade could not prevent damage. The sprawling new towns were not incorporated, there were no firefighters, and leadership given by the province in this area was practically nonexistent in the north. Fire rangers were undreamed of, other than those employed by private concerns.

The summer grew drier and hotter. There was no rain for weeks and July 10th saw record temperatures of 107°F. Only the wind gave some cooling relief. But there were small bush fires that day and the sky became red with flames that burned unchecked except by natural barriers of water and clearings. Some say that the section of fire which engulfed the gold camp started at Star Lake near Keefer Township, some thirty miles away. The thirty-acre slash that was South Porcupine had one main street, still littered with stumps, very few two-storey buildings, mainly log cabins, some still sporting tent roofs. There was one board sidewalk in the whole new community. Soon the infant settlement would be erased from the earth.

The threat of fire was obvious, and early in the day women and children were ferried from South Porcupine to Golden City. All available vessels — canoes, rowboats, even freight barges — were pressed into service, and still men had to wait their turn for a crossing on a day that seemed to grow progressively hotter by the minute. Meanwhile the small fires consolidated. One blaze, twenty miles wide in some places, roared in and engulfed the Porcupine, Cochrane and small communities in between. That cooling wind turned into a ninety-mile-an-hour gale, and a wall of flames said by observers to be up to 150 feet high in places rushed down upon forest, town and man. As opportunities for water rescue faded, some men waded into the waters of Porcupine Lake using anything that would float as a raft. When all other means of travel were gone, they waded into the water anyway. The outside world received its last message from the T. & N.O. dispatcher at 3:30 p.m.: "... fire has possession of the town ... very hot ... smoke so thick I can hardly read ... will close." The telegraph clacked to a stop and the operator rushed to safety.

A deserted South Porcupine? Not quite. Men still struggled to get to the water, and some died, engulfed in a white-hot fire. Tom Geddes, co-staker with George Bannerman, deliberately went back from the safety of the water to rescue his dog, and both man and beast were never seen again. A panic-stricken team of horses, somehow neglected in the press of the moment, rushed back into the teeth of the fire, and their death agonies were mercifully masked by the noise of the onrushing flames. Those in the lake covered their faces with wet blankets and periodically dipped their heads beneath the water. It must have been a strange sight — thick smoke, speeding fire and bobbing heads in the water. Those

Porcupine Muskeg
Golden City Ont.

*The Muskeg Special in 1911
carried both millionaires
and miners.*   – E. Audet

BREWER AND SHEPPARD
MEALS & BEDS

HILLS LANDING PORCUPINE
SMITH RIVER

A way station on the way to the Porcupine before the railway arrived, circa 1910.   – Timmins Museum

George Bannerman, one of the earliest mine finders, had a big camp.
– Author coln.

The only relaxed one is the lady. Caroline Maben Flower was both geologist and prospector.
– Timmins Museum

Prospector's shack. — OA-S15917

Finn miners from the Porcupine.
— Timmins Museum

Hunter Mine.
– Timmins Museum

Nothing remains of this Schumacher Mine today.

– Tomkinson

The Crown Chartered Mine in 1910 had a certain rustic charm, but it was a loser.
– R.W. Brock, Geological Survey of Canada, PA-45234

A prominent gold quartz vein at a new property. Tisdale Township 1910.
– Geological Survey of Canada, PA-45233

The Scottish Ontario Mine was one of the first mines in the Porcupines.   – Author coln.

Early assay shops lacked refinements.   – OAS8241

A diamond drill in operation at the West Dome Mine. – OA-S13743

The Vipond head frame was initially strictly utilitarian, Its size indicates the shaft was shallow.
Author coln.

115

The Dome Mine lost all power after the fire.
– OA-S13756

This mill was only meant for sampling but the stamps paid for Hollinger operation for one year.
– Ontario Department of Mines

Homeless by cook stove.
– Ontario Ministry Natural Resources

Shaft sinking in heavy quartz, Dome Mine. — PA-17433

117

The tents were soon followed by company housing.   – Bob Atkinson coln.

East Dome, South Porcupine.   – Bob Atkinson coln.

Newcomers found all the amenities of a bush town. — Author coln.

MAIN THOROUGHFARE, GOLDEN CITY, PORCUPINE

CITY HOTEL

The main street of Porcupine 1910. This was one year after discovery. Note Revillon Brothers store, the main competitor of the Hudson Bay Company.

— R. Brock, Geological Survey of Canada, PA-45326

First law office in the Por-cupine: J.J. Gray and Rev. Lawrence. – OA-S13707

The first school at Golden City, teacher Miss Williams, left back. Note absence of blackboards and insulation.

– OA-S15940

One fortunate asset for the area was electric power. In the Porcupine, electricity was ready as the mines required it; in contrast to Cobalt, where six years were to elapse before a general hydro-electric system was in operation. The Porcupine Power Company provided 3,000 horsepower in the early stages at Sandy Falls on the Mattagami River. Legend has it that, when Timmins came into being, the old-timers felt that civilization was getting too close and moved back into the bush. They must have been true bushmen, for the new town was a model of services compared to what people had enjoyed in the past. There should be no mistake, however, in a picture of the town at the time. There was a mixture of wooden shacks and brick buildings only gradually growing up to replace them. Roads were often ankle-deep in mud and horse buggies were the preferred means of travel. But with the advent of water mains and a sewer system, the rough-and-ready atmosphere soon faded.

The Porcupine was not immune to the labour unrest sweeping the rest of the country. A strike from November 1911 to June 1913 shut down much of the camp before outside workers were brought in to keep the mines open. With the advent of the railway, supplies became cheaper and the cost of living dropped. The mines tried to cut wages as a result and this focused the miners' grievances. The Ontario Provincial Police was called in to keep order. Although some men joined the One Big Union and paraded under its banner, the rhetoric soon died out and the men trickled back to work.

Examine the camp through the media. A visiting reporter form the Toronto *Globe* summed it all up in March 1912, " . . . the Porcupine Camp is passing from the explorative to the productive stage . . . no one has ever before received the benefit of intelligent investment of capital and expert management with which the Porcupine field has been favoured . . . ."

The local paper was *The Porcupine Advance*. From it we learn that Schumacher was practically a Yugoslav town, while Timmins was more European in its background. Mining was ever the main topic in the paper. Claims were advertised for sale on every page. The nickel newspaper ran an editorial praising the Timmins brothers. A lead story showed that the Hollinger had $10 million worth of ore blocked out at the 200-foot level. Panoramic pictures still bore witness to the new town's origins. Each one included tree stumps everywhere. Each edition had mention of mines and new prospects, many of which have faded from memory today. One wonders over the years, as people glanced at offers of men's felt hats direct from London for $2.50, accounts of Peary at the Pole and heavyweight Jack Johnson's prowess, if these "outside" events were strong competition to the daily mining quotations. Good citizens are ever-ready to complain about services, and in 1917 mining inspector J. Stovel said of roads outside the towns that, "They were not merely bad, they were damn bad."

The Porcupine was not a "poor man's camp" like Cobalt. In the silver centre the precious metal could be taken on surface or close to it for years. By contrast, once the initial surface gold had been quarried off, the Porcupine settled down to the regulated life of a hardrock mining camp at depth. The ore was middle- to low-grade in many places and large amounts of capital were required to recover the ore and mill it. The work force of 3,500 men in 1911 had to be paid and many housed and fed. The residence system was in effect. Average monthly wage was around $80 to $85 and for sixty cents a day full board was provided.

Maps of the time show the Porcupine area as a patchwork quilt of claims. There were few prospectors around. They had moved on to new rumours, new hopes of free gold and other precious metals. The prospector lives to find the stuff, and when a camp becomes fixed he moves on. Willet Miller visited in 1911 and remarked that little geological survey work had been done prior to that time. The mines were busy becoming established and the intervening great fire did not help serious ground exploration.

The three great mines made the camp, and as their progress is followed, the story of the Porcupine unfolds.

The early days of the McIntyre were difficult due to financing troubles. The mine lacked the guiding hand of a Timmins. It did have a good engineer, however. R.J. "Dick" Ennis opened the mine and stayed with it until his retirement in 1951. The story told that some early meetings were held in the central prison of New York City because a major officer of the mine was incarcerated there may be apocryphal, but many others seemingly just as extravagant are true. The effect of the 1912-13 miners' strike did not help the company, and whether the first gold brick poured was rushed to the bank to cover outstanding loans or not, many say that when it was deposited the gold was still warm.

At first the McIntyre operated on the side of Pearl Lake close to the highway, but later the plant was moved across the lake. There were problems with the quality of the ore. Hollinger veins dipped into the property but just as quickly dipped out again, so there was no relief. There were problems with the mill circuit and these took a long while to iron out. Five shafts were sunk before a good ore body was reached. Bad news travels fast in the mining fraternity. Shipments of materials began arriving marked cash on delivery, but there was no de la Mar to provide the hard cash on the spot. Frequently creditors called to find Ennis was underground and thus unable to meet with them. Any mine that offers the company doctor shares in payment of his account must have been in dire straits.

Prospects began to get better. Directors like Sir Henry Pellatt joined the firm. He had made his money in Cobalt and was well disposed to mining camps. The first dividend came in 1917 and the McIntyre began to consolidate its holdings. The Jupiter and Pearl Lake mines were acquired, and in 1924 the Plenaurum and Platt Veteran properties were added to the ground, making a total of 626 acres. Capitalization was now as high as $4 million. It was McIntyre money which was advanced to Conn Smythe when he built the Maple Leaf Gardens. The McIntyre was well established by 1920 and took its place as one of the three great producers of the camp.

To follow the progress of the Dome, go back to the time after the fire. Within two days of the disaster, directors journeyed north from such points as New York and Copper Cliff near Sudbury. From their private car, they finished the trip by handcar and then walked a mile into the mine. The destruction was surveyed and an emergency meeting gave the go ahead to rebuild. Those prompt visitors could reflect with satisfaction six months later on a new T. & N.O. spur line built to the mine, and in March 1912 the rebuilt mill was celebrated with a large banquet and smoker. This time twenty-one private cars came from New York and Chicago, and arrangements for the affair were in the hands of a Toronto firm, which even decorated the streets with appropriate bunting. The next morning the guests went out to the plant to see their gold mine in production. The roar of the forty-stamp mill was music to their ears. In the first year of operation the mill processed 400 tons a day and almost $1 million worth of gold was recovered.

Incorporated in 1912 with a working capital of $2,500,000, the Dome never looked back. The first year of operation provided a profit of $500,000. First Monell and then de la Mar became president. In 1916 the company acquired its neighbour, the Dome Extension, and added the latter's six claims to its own five. Over the next few years, $5 million in rich ore came from its open pits, while the mine covered its bets by proceeding to underground working. The war hampered production as it did with all the mines. At this time meals cost the Dome thirty-one cents, but the mine charged its work force a quarter, as men were hard to find. The new Workman's Compensation Act also increased costs, but that was a very necessary expense. With men at war the work force fell to the point where it was considered necessary to shut the mill and stockpile the ore until after the hostilities had ceased. This in itself was useful, for it enabled development work to be done. Jules Bache took over the mine in 1918 and ran it until 1942. The Dome was at last on a firm footing.

"Jewellery shop" is the expression most frequently used to describe the early showings at the Hollinger. After the great fire, torrential rain loosened much overburden and revealed even more quartz veins. The big mine started paying dividends only three years after it was incorporated, a staggering record when five to ten years was considered necessary to provide benefit for shareholders. The mill progressed from 500 tons a day to 800 in 1914. That year the Hollinger was able to announce it so far had fifty-four veins, with reserves estimated at a minimum of $13 million.

The mine was always a technological innovator. In 1916 electric locomotives were placed underground and paid for themselves in six months due to savings on the cost of tramming.

The future of the mine was assured when the company consolidated its assets and brought in the Acme and Millerton properties. At that time the two smaller properties had the most ore, but Noah Timmins felt certain that the Hollinger would surpass them in the end. A rectangular block of 440 acres was now the Hollinger Consolidated Gold Mines. Just after the war, many new houses were built for miners and other staff. The company provided a hospital and advanced working capital to various parties in the town of Timmins interested in going into retail businesses, so that there would be reasonable choice in shopping. The Hollinger had 2,600 shareholders and nearly all were resident in Canada.

In the twenties an estimate was made of the division of labour in the camp. New immigrants were often Italians and Russians, who did the shovelling and labouring jobs. Drillers could be Finns, Swedes, Austrians or Poles. Tradesmen were often men from the British Isles or Canadians, while the engineering positions were usually held by Canadians. This mixed group was joined in 1922 by one hundred Cornishmen that the Hollinger sponsored as experienced miners. Some of these newcomers might have worked in the Schumacher Mine, which Noah Timmins acquired for the Hollinger at close to $2 million.

Millionaire philanthropist Fred Schumacher's mine is finished but the community bears his name.
– OA-S13759

We can wonder today at company control of housing and other essentials of life, but consider the alternative. Accommodations in Timmins were usually either in boarding houses or shacks in the early days. M.E. Stortroen described such buildings as built on posts dug into the ground. A frame of two by fours was covered with inch lumber and tar paper on the outside and Beaverboard lined the walls. There was no insulation unless the builder was ahead of his time and put in some sawdust. Mining towns were not expected to last, went the reasoning, so why put up anything fancy?

Active mines on the Porcupine scene were the Coniaurum and Vipond. The Coniaurum was formed in 1924. An amalgamation of older properties, for a while it was under the control of the Coniagas Mine at Cobalt. The Vipond Mine had been started by Joe Vipond, one of the earliest prospectors in the Porcupine. These smaller mines operated in a camp where the veins often aided rapid expansion. This was due to their often vertical nature. When enclosed in firm rock, the veins were easily mined and at a less expensive rate, as they did not require heavy timbering to get at the ore. None of the solid producers were like an area property which has been nicknamed "the Spiritualist." This was so called because a medium apparently gave shaft sinking advice. There was a lot of money in the mine, but it all came from the surface, out of the pockets of investors.

In 1927 the McIntyre completed its great number eleven shaft, the large headframe of which is still seen across Pearl Lake. The six-compartment excavation was 4,250 feet deep and 160,000 tons of rock were removed in its construction, as well as 40,000 tons of water. The project included 304,000 feet of diamond drilling over 27,000 holes, 240,000 pounds of powder was set off to blast the rock, and 2,256,000 board feet of Douglas fir was used to timber the shaft. All mines at that time could reckon on one fifth of all operating costs spent on power.

Number 11 shaft house and waste ore bins, McIntyre Mine, 1929.   – PA-14336

In the same year as the McIntyre plunged deep into the ground, the Hollinger looked skyward and completed a three-and-a-half-mile aerial tramway. If progress appears only in such statistics, consider that in 1927 the Ferguson Highway joined the gold camp with Toronto. There was a great cavalcade of autos from northern points to celebrate the event, over a road which was only paved in the southern portions. One roadster called "South Porcupine Wild Cat" exists today in a photograph. When it completed the trip to the provincial capital, the car was dusty, dirty and sported quite a few dents and tire patches. The men leaning against it were all grinning. Timmins was on the map.

The gold camp was in the news in 1928 for a far less pleasant reason. On February 10th a fire began in an abandoned stope on the 550-foot level at the Hollinger Mine. Debris had accumulated there and the flames were easily fueled. Even after the call went to evacuate the underground workings, thirty-nine miners were found to be trapped below. At this time the industry was not equipped either by regulation or design to fight a major underground fire. The Department of Mines appealed to its United States counterpart and a relief train was sent north from Pennsylvania, carrying trained rescue personnel and equipment. The tragedy was that carbon monoxide had reached the miners and they were dead before help could arrive. The disaster had a sobering effect on the community, but out of it came hope for the future. The Mining Act was revamped as a direct result of the killer Hollinger fire and rescue training became compulsory in the industry. Timmins received one of the first mine rescue stations in 1929.

Nine coffins in a common grave, out of 39 lost February 1928 in the fire at the Hollinger Mine.
— Timmins Museum

Dick Ennis made a speech that year in which he remarked that the combined monthly payroll of the Porcupine mines was three-quarters of a million dollars. At his own mine, Sandy McIntyre received a pension for the use of his name. It was a good job that he did, for money never stuck in his pockets for long. The camp could afford to be expansive, for there was a total production of 176,626 ounces of gold and 932,732 ounces of silver in the year that closed the decade. A popular if irreverent saying of the time was that children prayed, "God bless Mummy, God bless Daddy, and God bless the Dome." That object of praise had its share of troubles. The mill was destroyed by fire but unexpectedly reaped a bonanza from misfortune. When the ashes were treated and the floors scraped, $500,000 worth of gold bullion was recovered.

Rich veins were always of interest. The 700 ft. level, Coniaurum Mine.
– PA-17521

130

It was good business to link a mine name with that of the district – Porcupine Vipond Mine, 1915.

– PA-30215

The Porcupine Premier Mine, 1916.

– OA-S15925

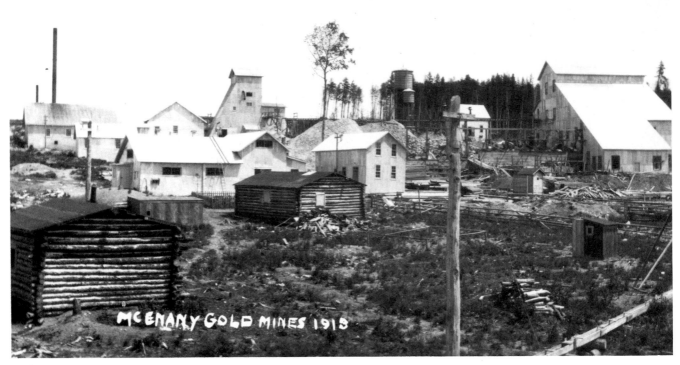

This Porcupine property never prospered. – PA-30042

Production needs came ahead of head frame cosmetics. Dome Lake Mine, South Porcupine 1913. – PA-30041

Three nation mine. — OA-S5649

Visitors going underground at the Schumacher Mine.
—OA-S13760

Gold is made ready for market in very ordinary surroundings. Gold pour at the Hollinger Mine.
— OA-S15925

Note the absence of hard hats or ear defenders in this early diamond drilling work at the Hollinger Mine.
— Author coln.

Early electric locomotive at Hollinger Mine.   – PA-17230

The filter press at the Coniaurum Mine mill, 1929. Only the foundations remain today.
– PA-14310

The government would take a dim view today of such activity close to residences.
– Bob Atkinson coln.

The Goldfields Hotel in Timmins is long gone despite its sturdy construction. – OA-15380-75

Sample of gold ore from Dome Mine, between the sixth and seventh levels. — PA-15214

In 1912, just three years after gold was discovered, the Porcupine was a big community with its own rail line. — OA-S13732

Miners coming off shift at the 1,800 foot level wait for the 'up' cage to surface. Hollinger Mine, 1936.

– PA-17570

Steady Production

The acceptability of substances other than gold for monetary
purposes, with the exception of silver, has . . . always been
confined within comparatively narrow limits in time and
space.

– Gold in Canada, 1933

The Dome mill was rebuilt in less than a year and the McIntyre completed a new 2,000-ton facility in the same period. The Hollinger was now mining more economically using the slice and fill method rather than that of shrinkage stopes. By this means ore was removed as mined and not left until the area was cleaned out. Successful mines are often a result of consolidation. Like the Buffalo-Ankerite, the Paymaster put two properties together to operate a fair-sized mine. Old-timers recalled that one of the shafts on the property was the site of the underground deaths during the 1911 fire. The Buffalo-Ankerite was incorporated in 1932 and held ground west of the Paymaster. Despite the Depression years, the camp had a high level of employment. Wages may have been low, but gold was a commodity always in demand.

The Dome was a mine reported to have been exhausted many times, but no one seemed to tell those working there. It is true that rich ore-bearing rock became mislaid once in a while. One commentator said that its ore bodies occurred erratically in the country rock, somewhat like plums in a plum pudding. In the Dome Extension, below the Dome twenty-third level, a rich ore body was discovered in 1933 and the mine continued to produce well. The Porcupine was a rich camp. The total value of all gold recovered from the area mines in 1932 was $22 million. Ontario produced three-quarters of all the gold mined in Canada, and ninety-six per cent of that came from the Porcupine and Kirkland Lake camps. In 1934 the office of the mining recorder was quietly moved from South Porcupine to Timmins, confirming the major role that town played in the development of the area.

McIntyre cage tender Jack O'Connell talked with labour activist Bob Miner in 1934. "Someday," he crystal-gazed, "miners will make a dollar an hour in this camp." Bob Miner retorted simply, "Jack, you've been smoking opium." But the country was awakening from the lost years and the price of gold leaped from $20 to $55 and everything picked up, including wages. The McIntyre built a well-appointed community centre and arena in the late thirties. Miners received $13 a month in coupons to spend at the complex. At the Dome, total production since 1917 to the end of 1937 was a healthy $75 million. The Hollinger commenced operating an electric train on surface between its number nineteen shaft house near Schumacher and the number eleven shaft. At twelve miles an hour, the trolley seated fifty men and beat the usual hike between working places, especially in miserable weather. The miners enjoyed the convenience of what they called "The Twentieth Century Limited" and, as the driver said, the outfit was a success because there was no competition.

Pamour Mine in Whitney Township was incorporated in 1936. Taken over later by Noranda, the new property became a big producer and put the focus back on the former

Golden City, now Porcupine, where many of its miners lived. The name Pamour is said originally to have honoured Wilfred D'Amour of Ottawa, who held some of the claims, but someone misspelled his name. The camp was at its peak with 1,258,670 ounces of silver produced in 1939 and 121,177,001 ounces of gold refined in the same year. The war intervened at that point and slowed mining everywhere in Canada. Labour became scarce and marginal mines like the Porcupine Crown and Vipond closed. The McIntyre had the same experience as other mines. One third of the work force, or 552 men, went away to fight for their country. Jules Basche, president of the Dome, arranged that its employees in the forces received a gift of $25 at Christmas. All area mines used their facilities to aid war production, and it was not until 1945 that the Porcupine camp really returned to its original line of business in strength.

The Hollinger surpassed all Porcupine mines in superlatives. In 1945 the mine found by a young barber-turned-prospector had reached a total dividend payout of $125 million. The mine had 350 miles of underground railway. Thirty-six electric locomotives capable of handling all the nearly 1,500 ore cars were maintained and operated by 165 men. The original principals had passed on, but their sons were active on the board of directors. Jules (son of Louis Timmins), Noah Timmins Junior, D.M. Dunlap (son of David), and Duncan McMartin's son Allen all carried on the family involvement in the Hollinger. As the Porcupine progressed to the late fifties, there were almost 6,000 miners working in its mines. The camp had just arrived at a total gold production of over $1.2 billion worth of the precious metal.

Eventually all great mining camps falter and fade. In 1968 the Hollinger closed down after almost sixty years of operation. The signs were not hard to read. In the early sixties the mill dropped in output from 8,000 tons per day to less than half that figure. In 1964 the 20,000th gold bar was poured, but over the next four years output trickled to a halt. The Hollinger brought out $566,200,241 in bullion before it closed. Typically, the company would survive in other mines it owned and diversity of investments. The McIntyre was never as rich a mine as the Hollinger, but it was the deepest property in the Porcupine and continues to operate to the present. Adapting to ore reserves, a 1,000-ton copper circuit was installed in the mill, and that base metal exceeded gold in annual production. Other smaller mines began to close. The Porcupine was beginning to lose its pride of place in Canadian mining camps. Throughout its history, new mines had come along at intervals, and in 1964 a great new discovery was made. A huge copper, zinc and silver mine was developed at Kidd Creek by the Texas Gulf Company. There was a familiar staking rush, some won and others lost their shirts, but that is another story.

The Hollinger gained a new lease on life in the seventies when it was reactivated for open pit mining. The big mine still has gold at depth and will get it out again when the price is right. As of 1970 the McIntyre had extracted $360 million worth of gold. No wonder Sandy McIntyre had regretted selling three quarters of his claims for $325.

The Dome entered the seventies using trackless diesel equipment underground and jumbo drills for greater efficiency. The Dome was a bigger mine with strong showings, but it was never as rich as the mine the Timmins brothers had started. By 1983 it had given up a total of almost 11 million ounces of gold and close to $2 million worth of silver. Just ten years before, the Porcupine had taken its own leap into the future. Thirty-five townships, encompassing approximately 1,260 square miles, including Porcupine, South Porcupine, Schumacher and Timmins, were amalgamated into the City of Timmins.

Today the Porcupine is bustling and as vital as ever. Over the life of the camp there have been an estimated total of 200 mines, and there will be more in future. Some never got past the name stage, while others poured money in but never took it out of the ground. Then there are the producers great and small that continue to contribute to the Canadian economy. Area maps are still crosshatched with properties, some famous and others equally obscure. Take a random sample from the news as these lines were written. Associated Porcupine thought it might reopen the Paymaster. Base metal Kidd Creek Mines was developing its own gold property at Hoyle Pond. Then in the next breath Falconbridge Mines were poised to take over Kidd Creek. Lastly, Jimberlana Minerals Limited of Australia was set to buy most of the other, older Porcupine properties, many of which had passed into the hands of Noranda Mines. Since the Aussie firm is a specialist in the recovery of gold from mine slimes or tailings, the area will be working on perhaps 200 million tons of the stuff lying waiting.

The future is bright for the Porcupine gold camp.

In 1960, assisted by Ed Hunt, President Jules R. Timmins, poured gold bar #18,490 bringing the Hollinger bullion production over $500,000,000 to date. — Timmins Museum

The aerial tram terminal and dumping sand used for backfill into steam locomotive cars. Hollinger Mine, 1936.
– PA-17596

A new tailings dam is under construction. Waste slurry or slimes will fill the site. Hollinger Mine, 1936.
– PA-17614

All this stock pile of B.C. fir square set timbers was destined for underground. Hollinger Mine, 1936.

A car of steaming timbers coming out of a tank of hot zinc chloride solution. Mine timber had to last a long time. Hollinger timber yards, 1936.
– PA-17607

143

Placing square set creosoted timbers into Preston East Dome's number 2 shaft in 1936. – PA-17725

The big mine had its own townsite. Dome Gold Mine, 1935. – Airmaps Ltd., PA-17662

144

A heavily timbered stope in the background is lit by miners' lamps. An extra post is ready to be put in place. 3,800 foot level, McIntyre Mine, 1936.
— PA-17569

Scaling down loose rock, 1,700 foot level. Hollinger Mine.   – C-22213

Dome miners having lunch in the main level close to the new shaft on the 2,200 foot level. Today, not too far away at Kidd Creek, there are underground dining rooms.   – C-22978

The drill repair shop near the #11 shaft of the McIntyre Mine in 1936 was bright and clean with whitewashed haydite walls.
– PA-17535

A train of ore cars is loaded using a power operated chute. Note the safety message. McIntyre Mine, 1936.
– PA-17567

After ore has passed through Hummer screens, it is conveyed to the rolls. The rock house at the McIntyre Mine, 1936.
                                                                                    – PA-17524

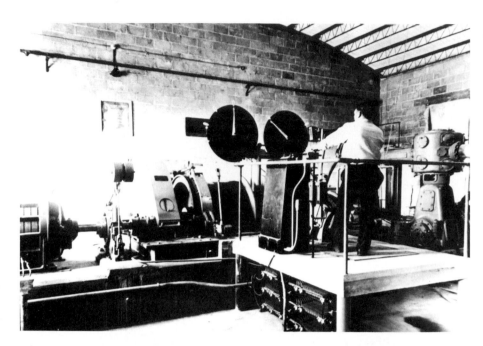

Hoist man at work, De Santis Mine, November 22, 1936.
– Timmins Museum

The new cage hoist for the shaft, with 10 foot drum and 700 H.P. Black indicator at the 2,200 foot level. Dome Mine, 1936.
– PA-17562

Pouring re-melted rough silver bars from a Monarch tilting furnace into moulds. Hollinger refinery, 1936.
– PA-17554

Weighing the refined gold brick at the Hollinger Mine, 1936.
– PA-17552

At the McIntyre Mine, men punched the clock 'out' before showering on their own time.

– PA-17583

The men have left. Their work clothes hang in the 'dry' ready for the next shift. McIntyre Mine, 1936. – PA-17581

Men coming off shift head for the punch house to clock off the job. Hollinger Mine, 1936.
– PA-17577

Sandy McIntyre found two great mines but he loved the bottle too much.   – OA-S11807

Porcupine People

I was there a blame not too soon. There's one thing about prospecting; you can never say you didn't have a chance.
— Reuben D'Aigle, 1959

I wonder if the world realizes what it owes, coming out of the Depression, to gold mining. It is the only business that has lasted since the first days of investment history; the only investment likely to last is a gold mine. In the Dome you have one.
— Jules S. Bache, 1936

There were so many characters and newsmakers who made the Porcupine camp. One character who impressed himself on early settlers was Jack Shields. He doubled as mobile snack bar operator and mailman. Newcomers to the Porcupine in the early years found that there was no food on the train north. If the traveller neglected to bring his own sandwiches, sustenance could be obtained from Shields. Tea was brewed in the baggage car and it was there that new pioneers could hear the mailman boast of his delivery prowess. At the head of steel, they were told, he could carry a full mailbag two miles before putting it down for a rest. One day this notion was put to the test, for when he set off to carry mail to settlers at the end of the track, a wag filled his pack with scrap metal. Jack caved in after a quarter of a mile of heavy packing. He never boasted of his strength again, but it is said he later took his own back on the prankster by serving up a dubious meal.

Jack Dalton ran the stage and livery service. There were no springs, but both passengers and freight reached their destination. As Dalton said, "With push and perseverance we could always swear our way through."

Caroline Maben Flower was a prospector. She missed out on her search for the golden rainbow but left behind the distinction of being the first woman in that hard-luck profession.

Unnamed but well remembered for their vital service are the convict labourers who improved the highways. Corduroy roads were laid over set sections of the trail. Their back-breaking labour had the bonus of being an effective deterrent to future crime.

One of the most famous pictures of the gold rush into the Porcupine shows eight men standing on the great outcropping at the Dome. Even in the black and white print, the thick gold vein flows almost like a liquid stream beneath their feet. In the foreground of the picture, a big, heavyset man stands astride the vein. His clothes are the thick woollen coat and pants, heavy boots and a wide-brimmed hat of a bush man. The most arresting feature is his full black beard, somewhat reminiscent of a man who has spent a lifetime in the outdoors. So it was for Father Paradis, missionary and prospector. He is said to have buried the first white man who died in the camp, but that was not his major claim to fame. Halfway on the overland route from Kelso to the Porcupine, he built a stopping place on the banks of the Fredrickhouse River, feeling that this could be an original ministry for travellers.

Father Paradis holds forth
– OA-S14643

Paradis built a house of unpeeled logs and hung out his shingle. For a steep $2 a night, those Porcupine bound could roll their blankets on his dirt floor or maybe use the soft side of a plank board. As an added attraction, a similar charge offered the right of winter storage for canoes. The host resisted the gold fever to serve the travellers, but the good father could not help but be touched by the mining talk that swirled around him. One day he was out canoeing when he saw a vein curving down through shoreside rock to dip below the level of the water. No one knows what prompted his next action. Quite possibly it was the prospect of making money to further his missionary work. Whatever the reason, the priest obtained dynamite, hand drilled the spot and fired a round at the vein.

In retrospect, we know that the vein was copper and not of any real worth. The location of the father's venture into prospecting was a natural bottleneck on the Fredrick-house. He unwittingly pulled the plug on the lake and the problem for future travellers was not the rate of water loss but the fact that the waterway was reduced to mud flats in short order. Paradis was roundly censured by those trying to pull a canoe across the muddy expanse. Paradis subsequently left the area but was pursued through the courts. The charge was interference with navigable waters. Like the travellers who floundered through the mud, the affair lingered until Paradis was absolved of any intent to interfere with the waters. A dam finally solved the problem, but the pioneering priest is recalled today largely for his unfortunate lapse into prospecting.

Another figure connected with water in the Porcupine was John Conlan. He provided the precious liquid. All mining camps needed a water man in the period between when newcomers relied on lake water and homes dictated a water system. At a nickel a pail, Conlan provided the basic necessities. He notched the doors of people who owed for delivery and had little trouble collecting past dues.

Other fondly remembered names were Foghorn Macdonald, late of Cobalt, Texas Steve and Bacon Rind King. One character ran a greenhouse. Called the White Rat, it was ostensibly a market garden operation but was in fact a bootlegger's headquarters. No one ever seems to have wondered about the large stream of people who were continually in search of "flowers and vegetables" despite the season.

Reuben D'Aigle is sometimes considered the forgotten man of the Porcupine. Fred Schumacher in this case would be a candidate for the least known. Schumacher made his money not in mining but in patent medicines. He was a wealthy man, but when he heard of the new gold strike in the Northern Ontario bush country, he decided to explore it for himself. In the early days the community between South Porcupine and Timmins was known as Aura Lake, for the precious metal beneath its surface. The name was changed to Schumacher with the connivance of Jake Englehart, the T. & N.O. chairman, after the 1911 fire.

Fred Schumacher's first association with the Porcupine was in 1912, and he brought hard cash with him. Within a short time he had purchased 160 acres of veteran land abutting the Dome for $8,000 and also eight acres between the McIntyre and Hollinger which were somehow overlooked in the general land rush. He left the big parcel and decided to concentrate on the eight acres.

He hired a crew, foreman and engineer and made ready to sink a shaft. But there was a problem. A house was situated right on top of the location for the proposed shaft. Mrs. Charles MacLean, the resident in that squatter's place, said later that Schumacher could have just moved the family off his property. Instead he purchased their house, and they were able to move elsewhere. That shaft was the only facility of Schumacher Mines. It was sold eventually to the Hollinger for just under $2 million. All the while he bought and sold property in the townsite, and it followed that the small community one mile east of Timmins should be named in his honour. Schumacher was but a visitor to the Porcupine. His residence was in Columbus, Ohio, where he built a great mansion and filled it with European art treasures.

The story of the twenty-four-year sale that Schumacher concluded can still bring a chuckle to older Porcupine residents. Shortly after he came into possession of the veteran's lots, he was offered $75,000 for the property near the Dome. Schumacher was in no hurry to sell and countered with an offer double the original. The Dome ignored the offer and the matter remained in abeyance for twenty years. But as the Dome prospered and attempted to expand, Schumacher was approached in 1931 and again offered the $150,000. The quiet American stated flatly that he would not bargain and that $150,000 was still his price. This time he pointed out that should his offer not be accepted, his price would be $300,000. The Dome president rejected the deal and in the next couple of years the underground workings approached the Schumacher property. This time the Dome president intervened, but Schumacher gently recalled his last price and indicated that if this sum was refused, the next round would see it at $600,000. Once more the project was dropped. A couple of years passed and then the Dome really needed the property they had so half-heartedly pursued for so many years. Facts available do not indicate if the Dome was "Schumachered" one more time. Something along those lines must have happened because the Dome finally received the land in 1936 for $1,125,000 and 20,000 Dome shares. In the lexicon of the Porcupine, "Schumachering" became a term for hard bargaining on a large scale.

Despite his stubborn approach in business, Schumacher became widely known in his adopted town for his philanthropies. In 1916 he gave money for Christmas gifts in his namesake community. A provision to continue this generous act was made in his will and the practice continues today. Typically, Fred Schumacher came and left the area without fanfare. Few people in Schumacher had met him. A photograph of Fred Schumacher shows a grey-haired old man with a warm smile directed straight at the camera. Schumacher Public School had a painting done in the thirties by Forbes with the same direct gaze. The local patron would be proud of his town today. For the people of Schumacher he will long be remembered with an epitaph once given by a good friend: "I never heard him say a bad word about anyone."

Remember Reuben D'Aigle, the man who came close to locating the Hollinger gold? An article appeared about him in the Toronto *Star* in 1959. Then eighty-five, D'Aigle was a tall, slim, erect man. Interviewed about his view of the Porcupine in the early days, he said, " . . . there was quartz all over the place but no more gold in it than you would find in a marble quarry." Despite his disappointment, the prospector said he did decide to go to Sudbury to record some of his claims. When he arrived at the nickel city, the recorder said he would have to go to Haileybury, and so he let it go. Reuben D'Aigle even staked iron in Labrador, but could not interest southern financiers. As he wryly put it, " . . . my claims there were like owning a mine on the moon." His daughter told me recently that all her father's papers are now in the archives at McGill University.

George Bannerman, the mine find-er who stayed on and built up the community.
– Timmins Museum

George Bannerman never did strike it rich in mining. Unlike most of the prospec-tors, he stayed in the Porcupine until his death in 1964 at the age of eighty-four. The teamster outfit Dalton ran became a bus line. Jack Wilson appeared solid-looking and practical with his high-cut boots, bashed-in hat and braces, but went broke speculating on the wheat market. Harry Preston, who stubbed his toe on a gold mine and later gave his name to the Preston East Dome, fell on hard times. In the early thirties, he appeared at the gates of the Dome for a free meal. Joseph Delamar, the financier who became Dome president and still paid cash for most things, died only a few years after the mine was established.

The men who staked the McIntyre had widely differing futures. Hans Buttner, said his son Fred, now resident in Columbus, Ohio, had a great deal more money than that which local lore credits him with receiving for his share in the rich McIntyre ground. He went back to Germany, bought a property for his family, travelled around the world and even learned English at Oxford. He also took boxing lessons there, and later came back to Haileybury and settled accounts with certain people who had badgered him about his short stature and called him "the Dutchman" instead of acknowledging his German nationality. He went to the United States and carved out a very respectable career for himself in

Hans Buttner, back left, with his father, sister and brothers back in Germany after co-discovering the McIntyre Mine, circa 1910-4.
– Timmins Museum

industry. In the thirties he came back to Canada, met up with his old partner briefly and even staked some claims which the family has retained. He died in Kalamazoo, Michigan, in 1972 at the ripe old age of eighty-seven. His son recalls that although his father had travelled widely and become a successful man, Hans Buttner's thoughts were always with the Porcupine, at least in spirit.

Sandy McIntyre, the man who gave his name to a great mine, lived in the public view for many years, but like so many other prospectors, he had trouble sticking to the money he earned. When he died in 1943 at the age of seventy-four, he was living in a cabin at Swastika with little means of support but making his way. Sandy's only recorded purchase with his newfound funds was said to be a spirited team of horses. He left the camp and went home to Scotland like many an emigrant before him, but life in the old country soon palled and he returned to Northern Ontario. He invested some money and lost it in a speculative scheme. Dame fortune smiled on him once more, but that adventure best belongs to the Kirkland Lake story. He received a pension from the McIntyre and would yarn with Dick Ennis when he picked up his cheque, but the ready wit and constant traveller faded to a shadow of his former self. He made the national news one more time when he had a chat in Gaelic with Canada's Governor-General, Lord Tweedsmuir, author John Buchan, when the latter passed through Swastika. McIntyre was then seen no more, except in Kirkland Lake.

One of the stakers of claims associated with the Hollinger, Tom Middleton, bears comparison with Sandy McIntyre. Jack Miller paid him $28,000 for his share, and he blew it in a few short years. A big, husky man in his prime during the Hollinger area staking, he eked out his last years on an old age pension, his health long since depleted, and he was partially crippled due to a leg amputated at the hip.

By contrast, Clary Dixon, the man in the canoe with Middleton on the day they heard Benny Hollinger's whoops of excitement, was more prudent. He took less cash but used the rest of his share to purchase 2,000 shares in the Hollinger. Youngest of the stakers of the Porcupine, he held on to his shares and they gave him a good income. He talked about his early life: "I bought some Dome shares at fifty cents and sold them at three dollars to pay for a couple of years study at university. You could say it was a rather expensive education because Dome kept going until in my time it reached forty dollars and even more." Then he said, "None of us, not even Noah Timmins, realized how important the camp was going to be, so we took the cash and let the credit go. The people who came later, when we made things easy, made the big money. But I'm afraid that's the way it will always be."

Alec Gillies did well for a while, but he went broke in some speculative venture. Benny Hollinger did better than any of the original stakers. His share amounted to $165,000 after he had given half the take to his original backer. He had interests in mining and other fields. A photograph taken of him in 1919 at the age of thirty-four shows a still youthful face. A quiet-mannered man with neatly combed black hair stares out of the old print. Benny never made another big strike and died a relatively young man of a heart attack. One prospector felt Jack Miller was the only one of the original stakers who not only did well out of the proceeds of his work but lived to enjoy them. He spent his winters in Florida and cruised the world.

Benny Hollinger, age 34, just before his death in 1919.
– Timmins Museum

What of the financiers who carried the Hollinger to producing reality? David Dunlap, who served the Timmins brothers well in Cobalt, was a strong partner in the big Porcupine mine. He greatly augmented his fortune there, left an estate of $6 million and gave enormous sums to charity. As well as the Toronto area observatory, the Donalda Experimental Farms near that city are lasting memorials to his generosity. The relationship that Dunlap had with the Timmins brothers was exceptional in the confidence that each placed in the other. Dunlap once made a statement remarkable for a lawyer: "In all our dealings there was never any written agreement between us. None was ever necessary."

Noah Timmins poured his energy and enthusiasm into the Hollinger and was its driving force for many years. Not content to rest on his laurels, he brought together the

At this first camp on the Hollinger property, 1909, Noah Timmins, second from right, has come to view his mine. Others left: Prospector ? Reid, Jim Labine, unknown, Alex Gillies and next to Timmins, R.G. Cambell, a survey party man.

Board meeting at the Hollinger. Left to right: Duncan McMartin, N.A. Timmins, Jr., Rt. Hon. C.D. Howe, Jules Timmins, President, A.A. McMartin, D.C. Finlay, J.A. Mc-Dougald, L.H. Timmins.

– Timmins Museum

principals of the Hollinger and the great new mining camp in Northeastern Quebec. The Hollinger interests advanced $3 million in venture capital to tide over the new Noranda in its growing pains. So one camp again bore the seeds of another. As things in mining ever come full circle, the Noranda companies were eventually to take over much of the Porcupine camp.

Noah Timmins was a quiet man and he was no publicity hound. Companies controlled by his organization ranged Canada and contributed in a large way to northern development. When he settled in Montreal, he became a leading figure in the financial community of that city. He never forgot the source of his great capital. In 1934, just two years before he died, he gave Fred Larose $15,000, a generous gift to the man who had sold the Timmins brothers the property in Cobalt which had given them their first big chance. His huge Tudor-style mansion, built in Montreal in 1930, was in the news in the sixties. Long passed from the family, its very size deterred buyers until an ingenious real estate agent had the place cut in half and sold as two dwellings. Noah's son Jules carried on the family tradition, first with Hollinger and then in the great iron ore deposits in Labrador. *Macleans* magazine has called Jules Timmins the "Shy Midas of Ungava." When Noah died at the age of sixty-nine, he had his monuments. They lie in the town which bears his name and in the prosperity of a community.

As you travel the streets of Timmins, consider some of the names. There are Gillies Street, Preston Lane, Middleton Avenue and Wilson Avenue, to name just a few. Some may feel that a street name is a poor monument, but it is enough if they are remembered. Men like these should not be forgotten. Their accomplishments do not show on the balance sheets, yet they opened up a large chunk of Northern Ontario.

THE TOWN THAT STANDS ON GOLD

Bill Wright: *Don't you sometimes worry, Harry, that this is all a dream, that it never really happened, that you will wake up one day and find yourself squatting over a tin plate of cold beans on some freezing scree in Alaska? I do.*

Harry Oakes: *I don't.*

There was hardly any more room than for the ore car.
– H. Grover, Author coln.

Examining the property.

– Gordon Peacock

168

A Country Awaiting Development

Gold is where you find it.

– Irritating old saying,
similar to that about silver.

I am a great believer in luck and I find the harder I work the more
I have of it.

– Stephen Leacock

The Precambrian era carved out the Canadian Shield and produced a much more complex work than even the most daring sculptor could conceive. Over seemingly endless time, there were great upheavals, folding of rock and still more violent forces changing the already misshapen land. Great volcanoes spewed up lava to cover the ground. There was wearing down, more folding, more fiery eruptions, and the cycle repeated itself. Meanwhile deep in the still-young earth, waters circulating at great temperatures began to find their way up fissures toward the surface. On the upward journey the waters cooled and deposited minerals on the fissure walls. When filled, the fissures formed veins, and patiently waited millions of years for discovery. In such a way were the gold veins of Kirkland Lake and the surrounding country formed.

The gold that lay beneath Kirkland Lake was in fine microscopic deposits within the veins. Only to the east did it show in concentration. Later it appeared the gold deposits were shallow, owing to surface finds which petered out after a while. It was ice that prepared the gold veins for their debut at upper levels. As the great ice streams ebbed and flowed, they laid some veins bare and hid others. Thousands of lake beds were hollowed out of the ancient rock. When the surface water later filled out these depressions to form lakes, they often covered mineral deposits. Then came mosses and lichens, vegetation appeared, and trees and bushes slowly gained a hold on the land. Poorly drained areas became swamp, and on steep rocks some trees surprisingly gained a footing in cracks in the bare country rock where the soil was practically nonexistent. Among all this the gold was randomly distributed without a care for location. The precious yellow metal lay in the rocks, sometimes in thick colourful belts and in other places in specks so fine as to be invisible to the naked eye. Under lakes covered by muskeg and swamp, overlaid with a tablecloth of mosses or bush humus, it waited for a breed of men persistent and curious enough to find it.

Pamphlets circulated by the railways and government agencies described the North for those anxious to leave the cities behind. In 1884 there was little to show on a map, so one writer settled for description. Settlers would find lakes teeming with fish, tracts of virgin timber and land ready for the taking. Cold weather was glossed over, swamps and rocks admittedly could be a barrier, but there was no mention of mineral wealth. Later, in 1897, an Ontario government pamphlet suggested that in the North " . . . independence could be achieved by a poor man sooner than the same position could be achieved elsewhere." Once again it was a life of primary industry that was depicted, and there appeared to be no possibility of manufacturing as enjoyed by the South. The pamphlets did not mention Kirkland Lake. The area was just white on the maps.

Surveyors started travelling north after 1900. Wealthy hunters also made forays into the bush in search of fish and game. Prospectors were scattered about the North, but they were solitary characters and did not talk much about what they saw anyway. Some prospectors found a huge deposit of iron in Boston Township in 1902, but iron was a glut on the market and the area was quickly forgotten.

While few people travelled in the North, others far away were about to make decisions which would lead them to this lonely land. In Western Canada, James Hughes was prospecting in the Yukon. Little is known of his success, but on arrival on the trail of '98 he found gold in a way different from the norm. He had a team of oxen and sold them on the hoof for eighty cents apiece. It was his first strike. His name would be made in another area. At the same time a young Chinese who lived and worked in Hong Kong decided to emigrate. He had no particular place in mind. Nor did a small Hungarian woman who was sightseeing in Paris. An Ontario government stenographer worked in Toronto and never did visit the Northeast. But all these people were soon to become attached to the north country.

The Temiskaming and Northern Ontario Railway began to be busy with the exciting developments at Cobalt. The line pushed north, locating small settlements in its wake. One such place was called Bell's Corner or Bell'sSiding. The place was north of Englehart and existed for a while as a contractor's camp. A small freight shed was erected there. Gold was seen on the tracks in the area of the camp, but the showings were small and no one took them seriously.

In the same year that the provincial railway crossed the Blanche River and founded Englehart, Joseph Burr Tyrell opened a mining engineering office in Toronto. Then forty-eight, he was a true link with Confederation. His father was an associate of John A. Macdonald and other founders of Canada. A doctor had suggested that he seek an outdoor life and would have been surprised to learn how literally his advice was taken. Tyrell joined the Geological Survey of Canada in 1880 and stated later, " . . . when I entered the survey all I knew was how to shoot . . . that came in very handy." When Tyrell finally left federal service and set out his shingle as an engineer, he had found the first source of oil in Edmonton, filled in great areas of the maps of Western Canada, located great coal deposits in Alberta and made important fossil discoveries. His acquaintance with the Northeast would come much later.

J.B. Tyrell, circa 1905-10. He explored Western Canada and then 'rescued' a mine in Kirkland Lake.   – E. Auckland

Cobalt made the southern financiers take notice and the wandering prospectors found the silver city was over-staked, so they moved north. There was more money for grubstaking now and several prospectors found themselves about three-days journey north of Cobalt, near a large lake not far from the Quebec border. Some gold was found near Larder Lake and this encouraged a larger crowd of optimists to follow. While searching the area around Larder Lake, they also ranged east and west, right to the little siding on the railway. They staked claims, but almost all of these lapsed for want of work.

That Larder Lake camp was a teaser. Surface gold was found but it went nowhere. The Toronto *Sunday World* of November 3, 1907, was impressed anyway and declared, " . . . the showing made throughout the district is a remarkable one . . . anyone can find samples of free gold on almost any one of the ore dumps without trouble." Heady stuff, but then the *Sunday World* didn't last very long either. The whole article sounded as if the reporter wrote his article in a bar on upper Yonge Street. One more experienced reporter set the record straight when he confessed no one could explain what the excitement was based on at Larder Lake. The whole area had to wait about thirty years before mines did develop, like the great Kerr Addison Mine at Virginiatown. The prospectors gradually drifted away looking for another area of promise.

The little siding north of Dane on the T. & N.O. was now dignified with the name Swastika. The community that grew there was destined to be the only one on the continent with that name. There were two small mines, one on either side of a small creek. The one on the west of the railway, on Otto Lake, was called the Swastika Mine, and the other within a short distance was called the Lucky Cross in an allusion to the community name. Both produced a little gold and had small mills. The Swastika, later called the Crescent Mine, was discovered by brothers Bill and Jim Dusty. One story has it that they saw a young lady wearing the crooked cross for a necklace and decided on the spot to take it for their mine. Another version tells that they saw the design on an Indian canoe, while yet another would have it that the idea came from the Egyptian symbol seen while they were leafing through an encyclopedia. No one knows for sure.

Two travellers passed through the area at this time. One was L.V. Rorke, a surveyor who was the assistant director of field surveys based in Toronto. When he arrived back at his desk after a lengthy field trip, his notes showed a large handsome lake three miles east of Swastika. Rorke had an efficient secretary and gave her an unexpected honour. He named the large lake Kirkland and a smaller lake Winnie. Thus did Winnie Kirkland give her maiden name to a lake and eventually a town which she was never destined to see. In later years her husband, Frank Joselin, explained that it was not an easy place to reach in those days, but although his wife never visited her namesake, she was always touched by the gesture. Another woman came much closer than a name. She passed by Swastika every so often as a cook on the train north to Cochrane. Recall Roza Brown, who made her debut in Cobalt. Now she passed along the line and made her plans. Unlike Winnie Kirkland, Roza Brown not only figured in the Kirkland Lake story, she went to live there.

Return for a while to J.B. Tyrell in Toronto. An unknown prospector likely picked Tyrell's name out of a directory with a pin because there were so many engineers. He sent him some grab samples of rock from a place called Porcupine Lake. Tyrell mislaid those pieces and later mixed them up with others from the Klondike. It was one of those grand opportunities so often missed in the early days of Canadian mining. When Tyrell realized his error he tried to option the claims from which the samples came. But by then it was too late and people like Noah Timmins did it first. Anyone slow off the mark in a promising new mining camp lives to rue it later. Not so with Tyrell, because he had another chance later.

Harry Oakes graduated with an arts degree in 1898 in his home state of Maine and even spent a couple of years in medical school. Then it occurred to him that doctors did not make as much money as he had hoped, and he left the university and went out to learn in the school of hard knocks. Harry was to spend a long apprenticeship, but he worked harder at it then most men. Any person who looks forward to leap years because they contain an extra day for working must have a future ahead of him.

William Wright had the same tenacity of purpose as Oakes when he became interested in something, but he gained his outlook on life in a different way. Born in the north of England in 1876, he was an apprentice butcher by the time he was fourteen. The army followed and he saw service in Egypt, India and South Africa. Bill Wright was one of those who survived the siege of Ladysmith in the Boer War. Ironically for a cavalry man who spent time in various Hussar regiments and later kept his own racehorse string, Wright ate horse meat to survive the siege. Wright was always a survivor. When someone asked him years later why he smoked so much, he said he acquired the habit to ease hunger pains. He ran out of smokes by the time the battle ended and recalled that an ample supply of cigarettes *then* beat all his triumphs later. He knew hunger and privation many times, but if he liked the military, Bill Wright was not an adventurer. He was quiet, temperate, unassuming, and even when he returned to the army he refused promotion. When he came to Canada he worked out West as a harvester before coming to Cobalt in 1908.

Somehow he did not fancy working in the mines and he teamed up with a fellow named Ed Hargreaves, who later became his brother-in-law. They worked in Haileybury painting houses. One woman, recalling the meagre living the two men made at painting, declared they were the hungriest painters she ever saw. Wright was bitten by the prospecting bug during his time in Haileybury. It could have come from stories he had heard of the Klondike or perhaps what he saw firsthand in and around Cobalt. Times were slow for painters and part-time help, and a man had time to dream. It took real effort to save with little income, but Wright was no stranger to discipline. Within six months he saved $40, bought a mining licence and then took the train north. He owned a veteran's lot in the Porcupine (given to soldiers after the Boer War), but lots in that part of the world had no reason to move yet, so he kept the deed and saved it for another time. For some reason he left the train and hiked into the bush at Swastika, toward the east, where people said gold had been found a few years before.

A mixed group rode the train northward in 1909 and 1910. Swastika was hardly noticed by settlers who had received government pamphlets extolling the virtues of the clay belt. They were Cochrane bound and hoped to farm. The would-be settlers sat next to a different breed of men. Maybe they were in the same coach as Benny Hollinger or any of the other prospectors who were to go on and find great mines only eighty miles ahead. Prospector Dennis Duffy put the philosophy of the prospectors into sharp focus. "There was the chance of making it big," he said. "If you struck it, you could make more in a day than a dozen years in other jobs. There was the life in the bush and the open air. And you were your own boss." There were no lottery tickets then; one made the chance by working at it.

The prospectors left the train around Driftwood, now Monteith, and packed three days into the Porcupine. Soon the news of a major series of strikes in the Porcupine drew all interested in mining to that area. But after a year, the prospectors began to get restless. Somewhere between Cobalt and the Porcupine there had to be more minerals. The two small mines at Swastika were a start, and gold had been found at Larder Lake. Who knew what might lay between? But the gold at Kirkland Lake was set deep and not so easy to locate as the Porcupine. As usual, only a few were lucky enough to find the elusive metal.

Many Finds Make a Camp

Where the gold was, I was not.
– J.E. Proulx from Porcupine, who slept on the
site of the future Teck Hughes Mine shaft and saw nothing.

*Gold has a mind of its own . . . gold is a woman. All the gold in
the world is waiting for just one thing, for the right man to find it
. . . . Maybe that is why I was never able to get my hands on it.*
– Roza Brown advised Harry Oakes
to try his luck east of Swastika.

The Temiskaming and Northern Ontario Railway was just about complete to the
Porcupine when Harry Oakes arrived by train in Swastika. To people like store owner
Frank Duncan, hotelkeeper Joe Boisvert and Roza Brown, who had settled down and was
running a laundry and bake shop, the newcomer looked like just another greenhorn in his
rumpled Palm Beach suit. Appearances do not make the man. Oakes may have been
dressed city fashion, but his face was lined and tanned by more than fifteen years of
outdoor life. He had a pugnacious jaw and a purposeful look. Roza Brown advised him to
go east if he were looking for gold, and within a short while he headed out in that direction.
The suit may have been new, but the pack and equipment bore the look of much use.

The train watchers had no way of knowing the so far chequered career of the
newcomer. A native of Maine, from a good middle-class family, Oakes went to Bowdoin
College, but his heart was not in more formal modes of making a living. Like so many
young men in the latter part of the nineteenth century, he had devoured newspaper reports
of the Yukon and Klondike rushes. Home and background cast aside, he took the long
route up the Inside Passage and over the mountain passes to the goldfields. Unlike so
many hopefuls, he actually did find placer gold, but the total count did not exceed $6,000.
By the time he had paid wages and fed himself, there was nothing left. Doctor Eugene
Whittridge, Oakes' hometown dentist in Foxcroft, Maine, chatted with the wanderer just
before he came north. He saw a man in his middle thirties with an intense way of looking
and talking, a man who could speak of practically nothing but his search for gold. Since
Alaska, where on one expedition he had even strayed across the Bering sea into Russia,
Oakes had followed his star all over the world. He had prospected in Death Valley, the
Belgian Congo, Australia and the Phillipines. At one point he had even farmed in New
Zealand, and had been both a hospital orderly and a ship's purser.

Harry had a purpose to everything. His visit to the dentist was no social call but
merely an effort to borrow money. The doctor was sympathetic but did not invest — and
was to regret it in a few short years. But there was always the Oakes family. His brother
was a lumber merchant and his sister Gertrude was in the United States government
service. Perhaps the outfit and way north came from them. Even so, Bill Wright, with his
small stake, was a prince compared to Oakes, who was almost broke when he arrived in
Swastika. If we are to believe Harry's recollection of his choice of a jumping off point, he
was influenced in his decision by a railway colonization pamphlet which he saw in
Arizona. He had noted the position of Swastika halfway between Cobalt and the
Porcupine.

Hotelkeeper Joe Boisvert had set up in Swastika because Guggenheim and Standard Oil geologists had told him there were promising mineral formations within a fifteen-mile radius of the new community. Jimmy Doig opened a hardware store. When he first arrived, his stock caused concern because some of the crates looked suspiciously like coffins. The people at Swastika and other villages on the line were soon to see real rough boxes when they went north for victims of the Porcupine fire. Swastika was a raw, muddy place in the spring of 1911. New settlers were miners from Cobalt. They were used to rough spots and the Lucky Cross and Swastika mines offered jobs. A photograph of the time shows five sober-faced men in city clothes holding hands outstretched across a veined rock perhaps twenty-five feet wide at the Swastika Mine. They were southern investors come, in the manner of the time, to inspect their faraway investment. How could these confident faces know that in two years both little mines would be closed?

For all its fragile nature, Swastika was to stay the larger centre for several years after the gold strikes at Kirkland Lake were made. Men with names like Orr, McKane, Reamsbottom, Maracle, McDougall, Matchett and Elliot are now largely forgotten. All passed from Swastika on their way east and staked some claims. They faded in time after the ground they took changed hands and became part of larger properties which were to make the third great mining camp within a range of 140 miles. Their claims were staked over a period of time, and the prospectors cannot be blamed if they sold out easily. Speculative money was not easy to land for an area without exciting surface finds.

George Minaker was one of those who stopped off at Swastika and walked the trail to Kirkland Lake. After the lack of success in the Larder Lake rush, Minaker had gone back to lumbering and teamster work. Maybe the news in the Porcupine stirred his old prospecting fever. He was eager to try again but was not exactly fighting competition from other prospectors when he staked three claims near Kirkland Lake. The day was February 23, 1911. The thirty-nine-year-old Minaker did not waste any time setting out his ground. The three claims extended south from a point just west of the present shopping mall on the edge of town. The claims ran in a north-south direction, and the most northerly would later be the key to a great mine. Minaker was spurred on by bitter cold weather, and he hiked back to Swastika and took the train to his base at Haileybury. A good lumberman could always find work. Others came and staked on either side of his claims. Curiously enough the claims directly to the east and north lapsed and were not staked until periods of four and eighteen months had passed respectively. The man that finally assembled them in one parcel was to make himself a fortune.

While the great fire raged through the Porcupine, Bill Wright was out prospecting around Kirkland Lake. He had been joined by his brother-in-law, Ed Hargreaves. The two men spent as much time hunting for food as they did looking for promising ground to stake. One story has it that the two were out rabbit hunting when they became separated. Hargreaves is said to have fired a shot to recall his partner. As Wright worked his way toward the sound, he came across a quartz outcropping. Although it was almost dusk, Wright saw free gold quite distinctly. It was in reddish feldspar porphyry. On July 26 three claims were staked. Wright did not realize his good fortune at the time, but two of these were directly on the main break or fault line of the area. That initial discovery was the first really rich find which established the Kirkland Lake camp. The next day Hargreaves went to Cobalt to borrow money from a mutual friend. The prospectors did not have enough cash to register their claims. Six weeks later Bill Wright staked another claim using his brother-in-law's licence, and later two more were staked on the future Sylvanite ground.

In the weeks that followed the staking of the four claims that were to become the great Wright-Hargreaves Mine, we see the difference in the characters of its two namesakes. Hargreaves had to feed a family. Following in the footsteps of countless prospectors the world over, both before and after him, he sold his share within a few short

Bill Wright before he became a wealthy mine owner.
– Museum of Northern History

weeks. The story goes that he received $6,000 for his equity, but it was likely less than that. He later lent his name to a small mine which amounted to nothing and largely fades from the Kirkland Lake story. The real luck Ed Hargreaves had was to marry Bill Wright's sister. She was a fine woman and they were to live in Barrie and later with Bill, until both men passed away in the early fifties. Bill Wright on the other hand, although he did not have the flamboyance of Harry Oakes, had the American's stubborn streak. He hung on to his claims despite days of poverty even more bitter than what he had experienced in Haileybury. His determination paid off in the end.

Joe Boisvert was one of those who noted it was one thing to have a rich mine prospect and another to have food for the table. "Those days were Bill's really tough days," he said later. "I can remember coming back from an assaying trip in December 1912 with a sled and a team of dogs. I decided to head for Bill's, as it was closing on evening. 'Just in time for dinner, Bill,' I called. His face went ashen and he didn't say a thing for a moment, then, 'Sorry, Joe, this is the wrong time, we haven't got a thing.' I unpacked the sled and we shared my grub. Bill and the man with him, Bill McDougall, really had it hard then — no tent, just rabbit skins to sleep in and a rabbit or the odd bird and nothing to smoke."

Late July 1911 Swift Burnside, a former varsity rugby captain, staked three claims to the east end of the camp on the same day Wright made his discovery. They would become part of a mine. There was no other land available between his claims and Wright's to the west. There was one person watching that area closely. Harry Oakes diligently pored over the claims registers at the mining recorder's office in Matheson whenever he could realize enough funds to buy a train ticket. His research paid off when he found that five claims north of the Burnside ground would become free for staking due to lack of assessment work. He kept this information and the time (after midnight on January 8, 1912) to himself. It was hard during the fall of 1911 to keep his knowledge secure. With enormous self-control he even waited until the evening of the seventh before trying his luck. He sat in the Doig store in Swastika yarning with George and Tom Tough. The two men were railway and haulage contractors. They had done some prospecting and Harry had an offer. If they were interested, action had to be taken that night.

He knew where the promising ground open for staking was located. He would split with the brothers if they paid the staking fee. A deal was struck there and then, and the trio set off in fifty-below-zero weather for the area four miles away. There was seven inches of fresh snow and they snowshoed off using a "bug" (a candle shielded in a can) whenever light for map reference was needed. Legend has it that each man wore three pairs of trousers to keep out the cold, but if they did Harry must have borrowed his, for at the time he had little but the clothes on his back. Eight hours passed from midnight before they had finished staking the five claims. As they lit a fire to warm up in those early hours of the morning, another figure loomed out of the early morning light. He also knew of the lapsed claims but was too late. Little did Oakes and the momentarily disgruntled Bill Wright realize that they soon would be partners in another venture. One story has it that Oakes went down to Gull Lake to get water for tea and fell in where a pressure ridge had weakened the ice. Wright and the others helped him out and he dried his clothes, dancing and singing as he cavorted before the flames. Harry Oakes had not seen the end of hard times, but his stake in this property would set him on the road to a fortune.

From that point on Oakes camped on the property in a tent and did not move until his prospects were again on the ascendant, then he moved to a small log cabin at the other end of the camp. The thick-set man with the penetrating brown eyes characteristically started working right away. He cleared trees and brush on the property, but little could be done until the snow was gone. Both he and the Toughs then systematically prospected their ground and found eleven veins in prophyry, greywacke and conglomerate. One of the veins was even found by George Tough when he sheltered under a tree during a rainstorm. A trench on the number two vein produced 101 tons ore-rich, ore-bearing rock, which yielded $46,221. The owners promptly invested this sum in a five-stamp mill. The Tough-Oakes Mine was to have the richest surface showings in the new camp and the shortest life. As for Harry Oakes, he soon lost interest in the day-to-day working of the mine that bore his name, because he had found another property.

For lumberman Walter Little, the discovery of gold opened up a line of work which was to one day take up all his energies. He agreed to do some hauling for the new mine. The idea made sense, for the small mines at Swastika also required freight shipments and there was a living in the business. Consider the route taken by the Tough-Oakes high-grade before the mill was installed. It was first hauled by stone boat or crude flat-bottom sleigh across a rough trail to Kirkland Lake. From there the bags of ore were loaded into canoes and floated over to the western edge of the lake. At this point the ore was loaded onto wagons for haulage to the train at Swastika. It was the type of operation which made citizens cry out for a road.

Back in February 1911 a prospector named Stephen Orr staked three claims, and in April John Reamsbottom staked three more. The six claims overlooked the west side of Kirkland Lake and had surface gold in places. A year later Jim Hughes, who had first found gold when he sold his oxen in the Klondike, arrived and bought the Reamsbottom claims. He hired a weatherworn Scot to prospect them. Recall Sandy McIntyre from the Porcupine story, the man who had joined with Hans Buttner to discover the great McIntyre mine. Profligate Sandy had spent all he received without thought for tomorrow. Actually he had not obtained much money for his share, being such a poor businessman when "the drink took me," as he candidly put it. Stories of the purchase of a team of black horses and even a trip overseas seem hard to accept on the strength of the money the McIntyre brought Sandy. A series of binges was far more likely the reason for his present flat-broke condition.

Hughes had the good fortune to buy one eighth of McIntyre's claims in the Por-cupine for $25. The sum seems incredible for the share in such a rich gold mine, which still operates today. There is no explanation other than the fact that Sandy's bargaining strength lay in inverse proportion to the supply of whisky on hand. There was no grudge

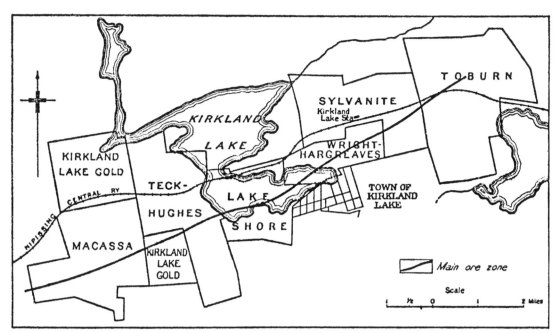

Kirkland Lake gold mines.

The gold camp right at the tracks at Swastika, 1910.

– H. Gadoury

177

borne by the bearded prospector when Jim Hughes hired him to prospect the newly acquired Reamsbottom claims. Mining is full of stories about events that might have been, and for the Scot each day brought a new opportunity. His partner Hans referred to McIntyre as a superb bushman when sober. He was a good choice to look over the property. Sandy tramped all over the ground and found some very promising veins. Hughes and McIntyre lined up the surface gold and felt that it was a link with Wright's vein on the east. It would remain for Harry Oakes to take the middle ground.

Three men staked three claims south of the Wright-Hargreaves property in a location covered today by a schoolyard, a post office and a library. Ed Horne, A. Maracle and Jack Matchett staked the claims at different times in 1911. When they were joined, the place became the Townsite Mine. It never did make any money, except when the Wright-Hargreaves worked its ground close to fifty years later. Ed Horne went on to greater things.

At the same time as this trio put in their stakes, C.A. McKane of Haileybury departed from his usual occupation. He was a plumber, but he had the prospecting bug and staked a claim on a hill west of the Hughes property. Dave Elliot was also out prospecting and he staked ground at the far west of the camp. It would be more than twenty years before it was developed further.

Joe Boisvert's hotel was the scene of a chance meeting in 1912. Bob Jowsey, who had done well in South Lorrain with a silver strike, was still in the mining game. He was now a promoter and he had a good chat with Al Wende, a Buffalo financier. Wende had made money at the Cripple Creek camp in Colorado and represented backers who were prepared to put money in promising new properties. The two were thrown together by a mutual dislike of onions. That was the odour which greeted them the next morning in the dining room of the hotel. They decided to forego this culinary pleasure and hiked off to Kirkland Lake. Bill Wright was cooking his breakfast. An invitation to eat is commonplace in the bush and Wright's offer of pancakes was more acceptable than the hotel fare. The meal turned out more profitable than either man realized. Wright told them of McKane's claim, which was now for sale. Jowsey thanked his host and elected to return to Haileybury at once to meet McKane. Wende passed up that opportunity in favour of exploring the prospects in the immediate area, and in the next few years he became a friend of Wright and Kirkland Lake. Bob Jowsey was able to purchase the plumber's claim for $3,500 and to pick up seven adjoining properties for $1,500 and some share consideration. In 1913 the amalgamated claims became the Kirkland Lake Gold Mine. It was the only one of the big mines to take the town name. This made for good advertising, but over a lengthy period the new company found that it takes more than a name to make a mine.

Oakes meanwhile never lost sight of his singleminded aim to make a strike and bring in his own mine. It was this obsession that shut out the hard existence he willingly endured. Forever short of funds — for no sooner did he receive money from the Tough-Oakes workings than he sank it into his new venture — Harry was grateful for any help. Roza Brown was one who gave him meals and was to champion him in future years when the abrasive Oakes personality wore down many associations from the early days. More than ten years after Oakes made his big strike, he talked about his life and philosophy as a prospector. He had four lean-to shelters and used these as bases somewhat as trappers do on a trap line. "I was up every morning before daybreak and on the go all day. At night I would head for the nearest of the little camps, cook myself a bite of supper and fall asleep dog tired. I worked hard, harder than a lot of these syndicated grubstaked prospectors think they have to do today.

"I am a great believer in a prospector working for himself and I think a district gets opened up faster if it has men working on their own or joined with one or two partners with the same line of thought.

178

" . . . I staked the Lakeshore for myself . . . there were old posts to be seen on it and the neighbouring Wright ground . . . it is a wonder that all the veins were not found.

"When I was there most of the prospectors seemed to be afraid of what they called the 'red granite.' It was too hard for them . . . they scattered off looking for soft schisted areas, but I recognized it for being really prophyry."

The property to which Oakes referred did not interest others easily. It was turned down for investment by several influential groups, including Noah Timmins and Sir Henry Pellatt. One of the reasons why the Kirkland Lake area was passed over by many prospectors was the general sparsity of free gold. Many dismissed the surface showings at the Tough-Oakes as being an exception to the rule. Oakes knew, however, that there were tellurides present as well as the prophyry. Tellurides are grains somewhat like the brassy pyrites but with a greenish colour. Harry had seen them in Australia, where their presence often signified high-grade ore. There was an old prospector's test for tellurides. An ore sample, when roasted, say on a wood stove, may produce a golden sweat as pin points of gold show on the surface.

A prospector named Melville McDougall had staked a claim near the lake at the same time as Oakes had first arrived. Oakes acquired that ground for a trifling sum and held on to it. He did not register the claim in his own name or admit to ownership of it for more than a year. Despite hard times, he held on to the claim, for he knew that one claim does not make a mine. In July 1912 two claims adjacent to the McDougall claim became open because of lack of assessment work. The two newly acquired claims were under Kirkland Lake. No one knows how Harry obtained the fourth claim. It was the one of George Minaker's four that he staked nearest to the lake. One thing is certain: Oakes did not pay much for it. Like so many of the prospectors, George Minaker did not have a lasting commitment to mining and he sold the one claim and was generally finished with Kirkland Lake. He had disposed of land worth a king's ransom and not realized the significance of the transaction. Now Oakes quietly transferred the McDougall and Minaker claims to his name. The four parcels fitted nicely together. He called the property Lakeshore.

Harry Oakes was a multi-million-aire when this picture was taken.
– Museum of Northern History

Harry Oakes had long since moved from the Tough-Oakes ground. He had built a tiny cabin on the Lakeshore, which he grandly called the Chateau. He saw Bill Wright frequently, for Wright had just staked a water claim that Oakes had not realized was up for staking, and the two men were connected from this time in their endeavours. They were of varying temperament, drawn together only by the search for gold. Harry Oakes, the man with the magnificent obsession, was living on the claims which would make him rich, but outward appearance belied that fact. The square-built man with the long jaw wore trousers of heavy mackinaw with wrinkled and baggy knees. The trousers were stuffed into high-legged boots. He often wore no top shirt but a heavy fleece-lined undershirt which stretched over sweat-stained braces. The man was no fashion plate but right in tune with standard work dress of his time.

Both Cobalt and the Porcupine were characterized by prompt exploration of promising discoveries. The Kirkland Lake area was not established as a full-fledged camp until seven years had passed. The initial lack of interest was hard on prospectors trying to hang on to their claims. A man had to eat while he performed assessment work, and one man working alone took a long time to meet his quota. As widely different in character as they were, Wright and Oakes had greater than average tenacity and held out while other men sold their ground in short order. Wright staked or bought a total of seventeen claims. Bill Wright financed his dreams by selling options on properties he presently did not need. Options bought time without loss of control. He made only one recorded error of judgment. He staked a water claim west of the Lakeshore but took in seven and a half acres too much. His ownership was disputed, and Wright lost the property which was sold to the Lakeshore Mine in 1915.

Harry Oakes had other claims as well. Part of the western end of the camp was his property, but it was twenty years before that real estate was to have promise of profit. In the meantime the Lakeshore ground occupied his working days. He hand steeled a two-compartment shaft commencing in the summer of 1912 and hired a man named Ernie Martin to work with him. Things were still tight. How else can purchases of two sticks of dynamite at fifty cents a time be explained? Only when money came in from the Tough-Oakes Mine did prospects look up. One day while Oakes was on a fundraising expedition to Haileybury, Martin found a quartz outcropping which had visible gold. The spot was named the number one vein, and on Harry's return the two men traced the vein on the south shore of the lake. It was at this point that Oakes decided to run a cross cut from his shaft to intersect what he felt had to be the vein on the main break. Jim McRae chatted with Oakes one day and was shown the newfound vein. They ate mulligan stew together while the mine owner talked persuasively about the mine he intended to bring in on the property. McRae came away with the impression of a proud man who had no doubt of his destiny in Canadian mining. He was right. The development of the Lakeshore and the other mines was just a matter of time.

The year before World War I, the new mining camp was a strung-out place. The main activity centred around the Tough-Oakes and Swastika, while everything else was slow in between. In mining, as in most enterprises, nothing succeeds like success, and the development of the Tough-Oakes Mine prompted both more exploration and promotion.

Roza Brown observed the newcomers. She watched the progress of the small mines and tramped around the area. Then without asking further counsel, the little five-foot-nothing middle-aged woman went out and staked some claims. Certainly no one thought it odd that she should try her hand at prospecting, and it was not recorded if anyone was surprised when she pitched a tent and lived on her claims for a year. One faded photograph survives, taken outside Roza's first house in Swastika. She stands with four men. Two were prospectors and one a landowner. All dwarf Roza, who appears in a long gingham dress, and man's peaked cap; the toes of her everpresent rubber boots show briefly beneath the hem of her dress. An apron indicates her calling as a cook, but the old print is kind and the apron appears white. Roza did not stand still long for anything but photographs. Within a year of camping out on her property, she sold it to a mining company. The rumoured price of $40,000 is probably high, but as usual Roza was not telling. Whatever she received was to be her stake. The sum bankrolled her future enterprises, but typically she kept on working, cooking in her little shack, boiling clothes and saving money. Today the Roza Brown claims still exist. Like hundreds of other properties, they never amounted to anything but the dreams of someone else. Roza had wisely taken the money while investors were still around.

After Sandy McIntyre located the gold at what became known as the Teck Hughes Mine, Jim Hughes incorporated the mine at a $2 million capitalization and had little difficulty attracting shareholders due to the fact that investors were already intrigued by the success of the Tough-Oakes Mine. Hughes was a fair man, and as public confidence grew in the venture he rewarded Sandy with a cash bonus and 150,000 shares. Unfortunately Sandy had not learned from his Porcupine misfortunes. He took a celebration trip to Montreal and when short of funds sold his shares for $4,500. In so doing, he frittered away what could have become a nest egg of $1.5 million. When asked what he had done with the money, the carefree Scot replied, "Why, I spent it on the drink." By contrast, a cook at the Teck Hughes later saved his money, put it into the mine's stock and wound up comfortably off. The business building he erected in downtown Kirkland Lake attests to his patience, a virtue sadly lacking in Sandy McIntyre.

Meanwhile the Tough-Oakes Mine had gone underground. An inclined shaft was serviced by a horse harnessed to a whim, or drum, attached to a cross shaft. The horse-drawn pulley brought up the ore bucket. This arrangement was not considered safe for men. The miners used ladders from surface. Oakes and his partners by now had made a deal with C.A. Foster of the Foster Mine in Cobalt. Foster assisted with financing the mine. He went to England to obtain the funds, and they were forthcoming through a British firm set up for his purpose and called the Kirkland Lake Proprietory Limited. Oakes gave up one third of his own interest in return for the share of financing arrangements and some cash. The cash was poured into his Lakeshore, but he was to regret the deal later.

Even the Wright-Hargreaves property was going through a period of depression. A promoter named Wendell Young had purchased Ed Hargreaves' interest, but he soon sold out to investors in Buffalo. A shallow shaft brought out three tons of rich ore, yielding 42 ounces of gold and 404 ounces of silver, but it petered out and work slowed almost to a standstill. If the Tough-Oakes production is not included, that was all the gold produced until 1917. No wonder Kirkland Lake was not well known on the stock exchanges of the time.

Harry Oakes was to say later that " . . . if Lakeshore had possessed a rich surface ore similar to that of the Tough-Oakes I would never had to go to the public for money." The irony in this was that Oakes had great difficulty in going to the public at all. Mining capital was still considered risky business by the conservative Toronto bankers, and private money was the answer. But the Toronto *Globe*, then the most influential newspaper, refused to take Lakeshore advertisements. There had been enough mining swindles and the paper did not want to be involved in anything but a sure thing. Nearer to home, the man who would one day be one of the richest men in Canada had trouble even getting credit for a pair of pants. Storekeeper Jimmy Doig refused him, feeling Harry's credit was exhausted. Oakes never forgot the episode and saw that his mine never did business with Doig.

So the mine owner pinched pennies and went without all but the bare necessities of life. How different his story was from the men in a similar position in Cobalt and the Porcupine! His future detractors may be confounded by two men who knew him then. Said George Cooper, "He'd give a fellow anything to keep his prospect going." Cooper should know, for he had enjoyed Oakes' hospitality on several occasions. If the new mine owner could not hire men, he worked more than one man's share himself. "Harry Oakes was just one of the boys at the time," recalled veteran prospector Dennis Duffy. "He was not very big, but he could work. He could drill as much with a single jack as two men could drill with a double jack." The old number one shaft at the Lakeshore was the one started by Oakes himself. How many big mining men can boast of having done the same thing? He put down the first thirty feet, drilling a round, blasting it off and mucking it out. There was a windlass arrangement and a wooden bucket. It was an endless round of drilling, blasting and mucking into the bucket. Then followed the climb to the surface, hauling the bucket and starting all over again. Oakes had guts and he worked hard. Times were always tough and teamster Walter Little often dropped off sandwiches from the cook at Tough-Oakes for the boys at Lakeshore.

Somehow the camp survived that hard year before the first war passed, despite a conviction among the mining fraternity that the gold only ran to shallow depths. An Englishman then living in Haileybury, Henry Cecil, bought up the Burnside claims and financed them through the same English firm that had provided the money for the Tough-Oakes. The Syvanite Mine was incorporated, as was Kirkland Lake Gold. The camp was pulling together into a recognizable unit, but as usual there were casualties. The tiny Swastika and Lucky Cross mines which had excited interest in the area at the start ran out of ore and closed down.

Another name for this early mine was the same as the village – Swastika. The men were likely shareholders up for a visit.
– OA-S5644

182

The Gold Camp Struggles to Grow

No more stock! No more stock!
> – Prospector Ernie Martin recalled Charlie Chow did not
> want to be paid for meals with share certificates.

Slowest camp I ever saw!
> – Remark by Harry Oakes to his dance partner,
> now Mrs. W.H. Cook

The Sylvanite Mine made its appearance in 1914. Developed from claims held by Bill Wright, it employed twenty men. The Lakeshore Mine was incorporated that year with 2 million shares, half of which stayed with the loyal Oakes family. Par value was ignored with the other million and these shares went from twenty-five to fifteen cents, depending on the market. The money raised enabled a work force of twenty-six men to operate and the original shaft was widened and driven down to depth. Joe Boisvert recalls that Harry was true to his friends in a backhanded way. At one point the stock shot up to fifty cents and he considered buying. Later he said, "Harry advised me against buying into it, telling me that the chances weren't good." Fortunately the investors in the market never heard that remark.

The Nipissing Mine of Cobalt optioned the Teck Hughes Mine, but Cobalt visitors were not enthused with the property by the edge of the lake. Only the Tough-Oakes showed real prospects. The rich gold-shot rock was still hand-picked and sent for complete processing elsewhere. In 1914, 213 tons of high-grade valued at $781,000 were shipped from the mine at the east of the camp. A transmission line from Charlton, twenty-six miles away, brought a much-needed 33,000 volts to the new camp. Plentiful power meant that the mine could commence a cyanide mill and be the first property to produce its gold from start to finish. Deep-level mining in Kirkland Lake was less expensive as the mines went underground, unlike in the Porcupine, as the rock was harder and the mines were on a smaller scale, but power costs were high for a long time and it was only as the Cobalt mines requirements lessened that electricity became more plentiful.

The gold camp was never short of high drama or men of interest. Henry Cecil occupied centre stage for a brief period. An English mining engineer representing powerful interests in Britain, Cecil visited the Kirkland Lake area, noted the proximity of the mines and compared their situation with other closely knit mining camps. British venture capital had earlier been invested in the camp at the Tough-Oakes and Burnside properties. Cecil cabled his backers and gave the opinion that at this early stage of its development the whole camp could be taken over with the right financial persuasion. He received approval for the scheme and quietly optioned several properties. Four engineers under Cecil's direction surveyed the area and then booked passage for England to give an on-the-spot report. They returned carrying reports, maps and ore samples. On the way through the Straits of Belle Isle their ship, the *Empress of Ireland*, wrecked and both men and reports were lost. The suggested amalgamation never took place and each mine was left to muddle along finding its own salvation.

Kirkland Lake Gold was reorganized in 1915. That is just a fancy way of saying the mine was having difficulties, specifically finding gold in regular values, and that more money was required if it was to carry on exploration work. Beaver Consolidated of Cobalt brought in their now disused plant for the property, but it was too big for the mine at the top of the hill and many technical difficulties resulted. One development from this activity was the adoption of the name for the hill. That sharp incline on the highway coming into Kirkland Lake from the west is still Beaver Hill, although the original outfit that held the property is long gone.

The Teck Hughes Mine now had a complement of forty-five men, but shares plummeted to five cents when the Nipissing option was dropped and it appeared that the mine would close. Abruptly those five-cent shares were all picked up and Handsome Charlie Dennison came on the scene. The nickname, as one may guess, was the reverse of Dennison's features, but it was a term of affection. The newcomer was heavily involved in the Buffalo Mine at the now fading Cobalt camp. He still regretted having not exercised an option on the Dome Mine in the Porcupine and had no intention of missing another opportunity, and so the Teck Hughes obtained a strong backer.

Wright's overstaking of his waterfront claim proved expensive for the Lakeshore Mine. That seven and a half extra acres cost the company $30,000 and 50,000 shares. The syndicate that owned the fraction received paper but had to wait a while for the money. In 1915 Bill Wright sold Lakeshore his claim for 200,000 shares, a directorship and vice-president's position. He was in no hurry to press for the cash.

Take a look now at the Lakeshore paybook for that period. Then as now there was a differential between surface and underground workers. Miners earned $3.75 a day, while surface labourers made an even $3. The fine copperplate entries indicate men of Scots, English and Irish origin. The melting pot of nationalities was to come later. All signed for their pay in a bold hand. Only a few signified the failure of formal education with a stubby 'x'.

An interesting character arrived in Kirkland Lake. Chang Kung Chow referred to himself as Charlie Chow. He opened a small eight-stool restaurant, and from the day the small place started it was always busy. Charlie had worked in many places since his arrival from Hong Kong in 1900 and had saved money for the store. The short, smiling Chinese never forgot how to save; this was demonstrated many times later. Fluent in his native

Charlie Chow who made it big in Kirkland Lake.
– Museum of Northern History

184

tongue, he never did bother to read and write in English, but he had no trouble counting in any language. The restaurant prospered because the food was good and Charlie was a friendly fellow. In the early tight-money days of the camp, men often received wages in stock. Charlie took shares at a discount of forty cents on the dollar. The restaurant owner amassed stacks of paper, some of which was only good for wallcovering, but other shares in time returned a huge investment.

Work began on the Elliot claims at the far west end of the camp. A shaft was sunk to about 500 feet, but the gold showings were poor. The property lay dormant there for ten more years until someone showed an interest in it. That group of claims would one day be the Macassa Mine, the only one operating today. Another reason why the camp languished at this time was the war overseas. Prospectors fighting for their country had their claims protected from loss due to lack of development work while they were away. New legislation in 1916 saw the loss of timber rights on those claims. Timber was considered a separate resource from mining. Harry Oakes had a different encounter over trees. When fire rangers approached the Lakeshore to cut down standing timber for a firebreak, he chased them off with an axe. Harry loved trees and the rangers felt their $2.50 a day less board was not worth the argument.

Andrew Farrell was working at the Tough-Oakes. He recalled a fire which swept in from the east. The people at the mine put their belongings in surface trenches. The fire passed without serious personal damage, but lack of ore soon put the mine out of commission more effectively than any natural disaster. Farrell earned fifteen cents an hour shovelling tailings through a flume. He frequently saw handfuls of small nuggets when work stopped to clean the ball mill. This would coincide with power interruptions whenever there were problems at the Charlton generating station. The Tough-Oakes dump was a favourite haunt for men off shift. Quite a few would pick up nuggets and have them set into watch chains. That gold on surface had been spectacular, but values dropped off as the mine went deeper. The mine milled 70,000 tons to the end of 1916 and recovered $1,385,000 in gold, but dividends ceased that year and the property struggled on for three more years.

Efforts to establish a mine at the Wright claims went on for four years. Then Al Wende took over as managing director and the mine was incorporated. There was an initial capitalization of 2.5 million shares and a good proportion of these were purchased by Wende's associates in Buffalo. A 150-ton mill was built, but it was called Wende's Folly by many in the camp, for the evidence so far did not justify a mill of that size. Such optimism in a mining camp was not new. The Rand Mine at the south end of the camp built a huge mill and never did find gold. By contrast, the Wende mill was a case of faith to be justified in a few short years. Bill Wright was not complaining anyway. He was the largest shareholder in the mine that bore his name. Now the shy, retiring man who had served his country once before went off to fight again. He was interviewed when he joined up and he emphasized then that Canada was his adopted home. He intended to come back. Stories circulated that he was the wealthiest private in the armed forces, but they were relative statements. The money was still on paper.

Harry Oakes was making headway at last. He had thirty-six men on the payroll and a good line of credit in town. He still looked more worker than mine owner. As money came in it disappeared underground to add to the Lakeshore finances. There were also legal fees as he took action to protect his investment in Tough-Oakes, for the settlement he had received still rankled. The Lakeshore was being explored at its third level and ore values began to slump. A horizontal drill hole was put out from that level to the north, but the results were disappointing. Any other man would have been overwhelmed at this point, but Oakes ordered a crosscut to be driven in that direction to meet the second vein he had found on surface. The cut was pushed through 300 feet of country rock without discovery

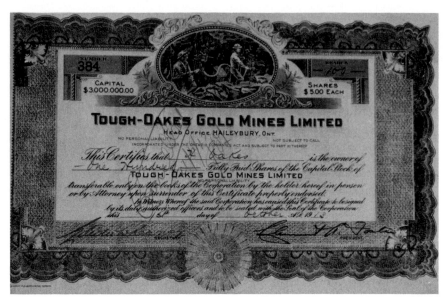

– Museum of Northern History

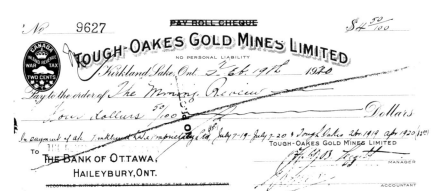

Note the fancy war tax stamp still in force after 2 years of peace.

The first tiny Chateau is seen on the left with the new Lakeshore Mine, right, 1913.

of note. Then came the shouts and excitement. The face showed gold everywhere. The long struggle was over.

Shift the scene now from the limited view cast by a miner's lamp to the sumptuous surroundings of a private railway car parked on a siding at Swastika. Observe Oakes meeting with Wetlaufer, Koons and other prominent investors from Buffalo and New York. The collective agreement on the table was for the purchase of half a million shares at thirty-two and a half cents each. Never again did Harry Oakes have to plead for money. As he sipped champagne and savoured his triumph, Harry must have reflected that his lack of success in the past few years had been to his advantage, for in clinging to those apparently worthless pieces of paper he was now rich and still had a large amount of stock. Harry owed this meeting to Al Wende, the newcomer who had persuaded his associates in the States to invest in the property. He rewarded Wende with 60,000 shares and a directorship in the company. Wende was to continue as a director for thirty-nine years.

Now that Lakeshore could erect a mill and have steady gold, the future of the community which had struggled along with its fledgling mines was assured. Some of the richest properties in Canada began to be developed at a proper rate. At last people could refrain from building houses with dynamite and gelignite boxes and concentrate on materials of more extravagant proportion. There is a popular story recounted in the area of a mysterious figure who visited the gold camp just as it was getting established. Local lore has it that Leon Trotsky was travelling across Canada to find out how men lived and he stopped off at Joe Boisvert's hotel and stayed a while. The attendant possibilities are rich. If this happened, did he talk to mining men who shortly before were broke and now were self-made millionaires? There is no answer because the story has never been substantiated. It just makes for a good yarn.

Stories that came out of tiny Boston Creek, just a few miles to the south, are more easily proved and none the less interesting. Take two episodes from many. The Barry Hollinger, a small mine which produced well for a while, had stock which sold in Canada for twenty-five cents but, because news was slower in reaching England, enriched mine coffers to the tune of sixty cents for the same paper. Sympathize with the Boston Kennedy, which had some gold but work ceased for good when the payroll was stolen. Today the small community has several abandoned workings and is under continual exploration with some promising developments. Long-time resident Desmond Woods mentioned a little-known claim to fame enjoyed by Boston Creek. Noted U.S. underworld figure Albert Anastasia, who was murdered in New Jersey in the fifties, had relatives by marriage there and visited frequently. The locals were interested in his bulletproof car.

Consider the state of the community. Electricity was no longer a luxury, for the new Northern Ontario Light and Power Company had arrived with power from Cobalt, and it was available for private rather than just mine use. There was no bank yet and mine pay came twice a month from Haileybury. Despite the absence of ready cash, a miner could still get a haircut for five Hargreaves shares. The main drag, Government Road, was still a sea of mud in wet weather. Vehicles passing the Teck Hughes Mine on their way into town had to ford a small stream. Most buildings were still tarpaper shacks, though often the log houses were of superior quality. Building lots were high at a thousand dollars. Al Wende found this steep, as well, and he had land surveyed in the east end for residential building and land prices soon tumbled. Charlie Chow was still having problems with too many payments in shares for meals, but this gradually eased off as money began to circulate freely.

By 1919 the struggle for survival was over. Kirkland Lake Gold had seventy-five men on the payroll, a new mill and its own water system for fire protection. Bill Wright was back to oversee his mine. Unlike the other Kirkland Lake mines, it closed from 1918 to 1920 to build a new headframe, mill and other buildings. The Teck Hughes had produced $67,000 in dividends and was for some time the biggest producer. It was milling its own ore and soon would have to upgrade the fifty-stamp mill. Lakeshore was at last paying a return on investor's money. The sixty-tons-per-day mill was going full blast and dividends exceeded $100,000. Harry's mother just did not believe the evidence of Lakeshore success when she had a dividend cheque in her hand. So many years of listening to her son's adventures and hopes were over. It remained to be seen if real fortune would now come his way.

As all mines expanded there was the old story of a shortage of skilled miners. In seven years the camp had come from a series of tiny mines strung along a muddy trail to a bustling small community which was rapidly taking roots. Then, with the perversity of life itself, there was a setback to the long-sought development. The first of two strikes in the camp's history took place. It dragged on from June 12 to October 18, 1919. The demand was for fifty cents more on a day's wages. There were no requests for safety aids or improvements in working conditions. No violence occurred. Men simply did not work. The mine whistles ceased to regulate town clocks and everything came to a standstill. No work was done except necessary pumping underground. When October arrived, the men went to work with no noticeable improvements in wages and the community settled down to steady growth.

By 1917 not all the land surrounding Kirkland Lake had been taken. There is no record as to the owner of the tame bear cub. — OA-15380-64

The La Belle Kirkland Mine included the camp name in its own to attract investors. This was a common practice but the mine was a dud. No trace remains today.
– Author coln.

This early view with the Wright Hargreaves Mine in the foreground shows the beauty of Kirkland Lake before slime dumping destroyed it.
– OA-13563-64

An Ontario government liquor store stands today on the site of one of the Wright Hargreaves head frames. It is a different type of gold mine!

– OA-S17951

190

Steady Development

It's a man's duty to live as long as he can.
   – Dr. Joseph B. Tyrell, the greatest land explorer of his day,
      came to be the manager of the Kirkland Lake Gold Mine
         and also lived to a ripe old age.

Quite a large interest in the camp is held by the original prospectors, whose interest still seems to be in the profit from the ore rather than the other fellow's pocket in the stock market.
   – Harry Oakes, 1920, did not care much for speculators.

Within the four miles by three quarters of a mile which became the Kirkland Lake camp, the community was now organized as the Township of Teck. Today that name is carried on in one of Canada's great mining companies. One local need was for an improved road link between Swastika and Kirkland Lake. The municipal body also petitioned for free assays for prospectors at the just-opened mining recorder's office in Swastika. New Protestant pastor James Lyttle had obtained a building lot, donated by Harry Oakes, high up on the rocks overlooking Government Road. The mines were generous in church-building support. A cancelled cheque for $100 for the new church still survives. Drawn through the Haileybury branch of the Royal Bank, it was a gift from the Tough-Oakes Mine, which even in charity proclaimed proudly that it had "no personal liability." That same cheque, on display at the present United Church, has a red embossed Canada Inland War Revenue two-cent stamp, reminding us that although wars end, taxes are ever with us.

Kirkland Lake was never short of characters. Arthur Slaght, a barrister from Haileybury and solicitor for the Lakeshore Mine, centred his individuality on his dog, Cobalt. He was the dog that owned a district. Cobalt was a bulldog with the ugliest face in a breed not known for good looks, but his friendly nature won the affection of everyone. Although his main base was Cobalt, he would often journey up and down the T. & N.O. line or on the lake boats as his fancy dictated. When Slaght was in court, Cobalt often attended to add a little dignity to the proceedings. Hotelkeepers usually entered "Cobalt Slaght, Basement" in their registers when he honoured their establishments with his presence. He also visited the gold camp and left his stamp of approval.

All sorts of people made their homes in the town, and a variety of nicknames charmed the ear. George Bernard Shaw was considered too cumbersome and Shaw became plain Casey Cobalt. A big man from middle Europe was christened the Russian Kid and the name stuck. Any set of unproven claims had a special name. They were universally called "prospects." One definition of prospect means hope, and for many this was all their property became, a lottery ticket on the bounty of nature. The town's biggest character moved in from Swastika in 1920. Roza Brown had a rocky lot facing the lake and lived in a shack which housed her laundry and had a sign which urged "Everybody deal at Rosies." Behind the house were other small shacks owned by the hard-working pioneer. They were rented by a variety of young ladies with no visible means of support. Everyone was welcome in Kirkland Lake.

The North has always been an interesting area for crime and the occasional frustration of the law. Liquor offences plagued the judicial process, but on one occasion the scales of justice were tipped by the magistrate himself. A common practice of the Attorney General's department was to employ spotters or plain-clothes investigators of liquor offences. One enterprising spotter disguised himself as a returning officer and travelled the railway. One day he faked a heart attack and lay gasping for a little stimulation. The passengers were concerned and one good samaritan tendered his hip flask. The "soldier" immediately recovered, arrested the compassionate one and took him off the train at Swastika to the accompaniment of catcalls and other unkind references from passengers. When the case came up before the Reeve, Major E.S. Gordon, who was justice of the peace by virtue of his office, the spotter was confounded. The magistrate fined him $200 or six months in jail for impersonating an officer. The citizen by contrast was fined $10 and warned to be more careful in his choice of recipients for aid. The court later received a polite letter from the Attorney General thanking it for the dispensation of justice. The letter suggested, however, that in future a full-time stipendary magistrate would be appointed. As today, originality in sentencing was not appreciated in Toronto, especially where public servants were involved.

By 1921 the Wright-Hargreaves mill was operating at one hundred tons a day and produced $458,751 in gold. The Teck Hughes capped production with 160 tons. Harry Oakes won the prolonged litigation over his interest in the Tough-Oakes Mine, but the award hardly paid his legal fees. Satisfaction was in the much more rich fruit of victory, and he had the judgment printed and distributed to his friends. Comfortable in the knowledge that he had controlling interest in the Lakeshore Mine, Oakes could at last indulge such fancies. A popular writer has described him as " . . . assertive and self-made, full of defiance and conviction." Surely this was an apt description for a man who sacrificed everything on a gamble that finally paid off and who had worked hard for all he now enjoyed.

Newcomers were commonplace in the early twenties. George Wyatt arrived April 10, 1921. He went to the Wright-Hargreaves Mine, as he heard a fellow Cornishman might be working there. A man he met on the property chatted with him about rural life. Wyatt had been a farmer and the man he later knew as Al Wende offered him a job on the spot. Wyatt went underground the next day. He donned the spluttering carbide lamp and worked below surface for a few months. At that time a miner received $4.75 a day, a helper $4.25 and surface labourers took home $3.75. One day Wyatt was called to surface. Could he kill pigs, enquired Wende? The mine had its own bunkhouse and cookery. Livestock was kept, but no one was knowledgeable in meat preparation. Wyatt could do the job and maybe this expertise led to the chance to try surface hoisting. He stayed at that work for forty-two years, just one example of many who were in local mining a long time. He remembered Bill Wright as a quiet, reserved man, a pleasant conversationalist who especially enjoyed talking about horses. Harry Oakes would ski across the lake for a chat in the hoist room on a Sunday when the town was still. The hoistman recalls Oakes as having little to talk about other than mining.

Charlie Chow's restaurant continued to be the popular eating spot. Unfortunately for some patrons it was also the centre of activity not normally found in a restaurant. Like many storekeepers, Charlie kept stove wood stacked in the street. An old-timer recalls the occasional prank of sliding logs across the ice into the restaurant and Charlie just as promptly bowling them back into the street. Fights often started in the restaurant and losers found themselves out in the street in short order. The Russian Kid often fell into such scrapes. He wore a high-domed Stetson hat. One wonders if he were not better named the "rushing kid," as stories about him seem to linger all over town. Charlie was not amused when such horseplay occurred. When the stove was knocked over, he frantically shovelled burning coals into the street to save the place from burning down.

The great mill and #1 head frame of the Lakeshore Mine are gone now. The spot is covered by a shopping mall.
— Geological Survey of Canada, 1929, OA-S17950

Ore loading pockets at the 2,200 foot level, Lakeshore Mine.
— Author coln.

The Sylvanite Mine, 1926. — PA-15518

The year 1922 offers different memories. For some it was the year work began in earnest on the Sylvanite Mine. For others it was the beginning of petitions for a railway branch to run from Swastika to Kirkland Lake and the mining areas to the east. It was also the year that the mines ringing Kirkland Lake started to dump mine slimes into the lake. No one is recorded as having protested the act, and from that time a beautiful waterway had but a few years left. Even Harry Oakes lived in his lakeside Chateau and condoned the action. Lakeshore Mine was a big plant operating only a stone's throw from the building. With all that noise and activity, slime dumping was just part of the scenery. Today the setting of Oakes' former home is both rustic and peaceful.

There had been two major fires in the North. The Porcupine fire of 1911 and the Matheson fire of 1916 were still well remembered when a power failure in Kirkland Lake in October 1922 gave warning of yet another disaster, the Haileybury fire. Kirkland Lake was far removed from the horror of a giant bush fire. One of the reasons was that the local volunteer department always went out to fight bush fires whenever they were sighted. Ed Hargreaves' wife and daughter were travelling north from Toronto when they heard that the fire had wiped out Haileybury. They continued to North Bay and were overjoyed that the rest of the family was waiting to meet them. Ed's brother-in-law, mine owner Bill Wright, lost his big house in Haileybury and he moved to Barrie and took up permanent residence there.

One prospector, Jerry Shay, sold his claims near Swastika to an Englishman, David Freeman Mitford. The new owner, later to become Lord Redesdale, entered into the spirit of things and lived on his property for several years. The place was what people in the mining fraternity called a "no good property" and Redesdale is better known for his daughters than for his mining success. The girls visited in summer months. As they grew their individuality asserted itself. Unity and Diana became enamoured of the fascist movement. Unity was a great friend and admirer of Adolf Hitler. She shot herself after the

Nazi leader spurned her when he came to power and lingered for a while because she botched the job. Diana married Sir Oswald Mosley, leader of Britain's fascists. Nancy became a writer and her books have been read widely. She became famous for her book *Noblesse Oblige*, a 1956 account of "U and Non-U" or upper- and non-upper-class use of language. One Swastika story is that Hitler met Redesdale in 1923, admired the Swastika emblem that the latter had on his watch chain and adopted it for his own symbol. The truth is probably quite different.

Lord Redesdale, September 14, 1923. — Museum of Northern History

Biggest producer of the mines in 1923 was the Teck Hughes. It had a wealth of ore and grossed over a million dollars, the first mine in the camp to pass that figure, and was averaging profits of $60,000 a month. The mine at that point was the deepest in the camp at 1,105 feet. Life on the property was not all uneventful. Only two years before, surface workers had been startled to see strangers setting up claim posts on the ground. A clerk had forgotten to remit taxes due and the mine site was thrown open for staking. Certain local speculators started to claim the ground, but the Minister of Mines relented after several urgent calls came from the mine. The scavengers were called off and the Teck Hughes Mine never suffered any comparable misfortune again.

The Lakeshore Mine was treating a hundred tons a day in 1923. There were no reserves built up and the ore came from development drifts, but the gold was there. At the Wright-Hargreaves Mine shaft sinking was in progress to the thousand-foot level. The going was expensive, but a few stopes were up to forty feet wide and some savings were realized as a result. The once-spectacular Tough-Oakes Mine and the neighbouring Burnside were amalgamated in 1923. The 343 acres had a rich surface but gave disappointing results underground. Certain veins merged and the Burnside shaft was deepened to a thousand feet. From the days when unkind cuts were made on the London stock exchange about the "Tough Hoax," the newly joined property settled down from the heady days of surface high-grade to a modest production at depth. Now it had a new nickname in Canada. A customer in the broker's office was understood when he referred to the "Tough Old Bloke."

Town population stood at 1,400 and with its haphazard appearance, housing being governed by the location of rocks, the place seemed to be bigger than its numbers

Harry Oakes sits on the left at the Lakeshore bunk house, 1922.
– Museum of Northern History

The Lakeshore Mine with the bunk house foreground. The men's quarters is now a motel.
– Museum of Northern History

indicated. One local spot that also prospered was the Kirkland Lake Hotel. Nicknamed the Ashcan for its siding fabricated from cyanide cans used in the local mine mills, and its owner, Jimmy Ash, the building had been brought up from Cobalt in sections. Visitors to the hotel on Dominion Day could attend water sports on the parts of Kirkland Lake not spoiled by the encroaching slimes. There were field sports on schoolgrounds and mining events took place at the Wright-Hargreaves Mine. A photograph of the time depicts the machine-drilling event. There was a large crowd of serious men gathered around the machines. Reputations depended on the skills being demonstrated.

The Tough-Oakes-Burnside Mine caught fire and burned early one morning in May 1924. Although the fire brigade arrived and attached hoses to mine hydrants, the staff house, office and one house were destroyed at a loss in excess of $20,000. Men just off shift who were sleeping in the staff house were forced to jump from a second-storey window. All that was saved was the concrete vault. The small building contained documents and several gold bars, and it was several days before the steel door cooled enough to be opened. That vault can be seen today on the highway at the east end of Kirkland Lake. It is all that remains of the office and an aerial tramway which crossed at this point.

Charlie Chow opened a new hotel. It offered the most modern facilities for dining and accommodation in town. Charlie was lucky to see it open. While the new building was still under construction, he was sweeping snow from the lower roof when a slide took place, carrying him with it. Charlie wound up on the ground with nasty cuts and bruises. A glance at the hotel registers from that time show the place was frequented by a clientele from all walks of life. There are blotters with Chinese advertisements tucked in the pages. A bill is made out on the back of an Eaton's envelope at three meals for $1.50. One order to Gamble Robinson, wholesalers in Cobalt, has a brief note dictated by Charlie. "As to my financial standing, I wish to refer you to the Imperial Bank here." Did anyone ever bother to check Charlie's credit rating? Any clerk would probably have been told that he had enough money to buy out the wholesaler.

In the summer of 1924 the Rev. Cyril Goodier was the first Anglican priest in charge of the new mission. One day, while out on a walk, he met two men patching a canoe in a garden at the Lakeshore Mine. One said he was an Episcopalian and might be able to help with land for a new church. After he left them, Goodier realized that the potential benefactor was Harry Oakes. The mine owner met Goodier again and took the priest to a rocky lot overlooking the schoolgrounds. "This is the best place, the church will stand high and nothing can be built in front of it," said Oakes. The lot was not sold but rented by the year. So the new St. Peter's on the Rock was started. The deed for the ground came after Oakes had finally left town.

Harry Oakes relaxed his business interests in 1924, for he had other matters on his mind. He had time to give one and a half acres of land to match that already held by the new hospital board for building purposes and took all $15,000 worth of debentures offered for the projected new building. He became a citizen and took a world cruise. He met a girl half his age on the voyage. Eunice McIntyre was twenty-four, the daughter of an Australian civil servant. There was a short courtship, they married and had twenty years together. Peter Newman has described Harry Oakes as a short, belligerent barrel of a man with a nose shaped like a half-empty toothpaste tube and eyes that appeared to shift in an instant from a man's expression to the motives behind it. If Oakes had a vice it was work, for he drank little and the bronchitis that plagued him precluded smoking. He could be generous to a fault, but only if the act was his own. His brother Louis and sister Gertrude became rich because they stuck with him through the lean years. Gertrude was a working director and was busy with Lakeshore affairs right up to the year that she died. Ernie Martin, who worked with him in the original two-man operation, received shares worth

$100,000 for his loyalty. Yet Harry would not give money to the hobo who stopped him for the price of a cup of coffee on Government Road. "Go to the Lakeshore gate if you want a job" was his brusque answer.

At a time when the unusual was not commented upon — since the usual was in process of being defined — Miss Mabel Fetterly stands out as a pioneer woman. She was a prospector and staked claims near the Upper Canada Mine. Between prospecting trips, she drove a transport team the eighteen miles from Dane to the Argonaut Mine near Larder Lake. Considering that a team consisted of nine horses, Mabel Fetterly had no easy job. Her tent was pitched by the railway tracks at Swastika for years and she cut wood for sale as additional income. The opposite of a working teamster, Cobalt Dick Elliot drove his team for pleasure. He wore a handsome coonskin cap and drove a smart pair of horses hitched to an elegant cutter. In town, that engaging character Charlie Chow was going through his Garbo stage. Frequently he stood behind the counter and solemnly told newcomers that Charlie was out. There was no rhyme or reason to the deception, just Charlie's idea of a joke. Twenty years later he would hint darkly to children that he was Charlie Chan, the famous detective.

Someone evidently disliked dogs in the camp. The Wright-Hargreaves offered $100 for information leading to the arrest of persons who shot the mine bulldog. Bill Wright was not around to find out the success of the advertisement. He made a trip to England and transacted some business at the same time. He still owned a veteran's lot in the Porcupine, but the original deed had not been transferred. A cable to the old friend who held the deed did not produce any satisfaction because the latter had heard of the Porcupine gold camp and would not do business by letter. Wright visited him in London and obtained transfer for $20,000. The sum seems exorbitant for a plot of land that had originally changed hands for $10. Wright made his profit twenty years later, however, when in 1945 the Buffalo Ankerite Mine purchased the lot.

Sandy McIntyre was prospecting in the Red Lake area in the mid-twenties. He had seen service in World War I, but army discipline had not suited him and he drifted to a timber-cutting squad in Scotland. Their nickname, "Cold Footed Thumb Suckers," gives a strong indication that this was an outfit for military misfits. Always articulate and a good companion in the bush, he was ruined by drink when he reached the city. Showing an urban dweller how to set a campfire by using a hotel room chair for kindling, right in the room, is just one example of the scrapes he indulged in when "the drink had me." Sandy could tell witty stories, like the one where he found a bear foraging for food, wedged half in and half out of his cabin. As Sandy told it, he shot the animal, and since he was too heavy to move, took his axe and whittled the bruin away in stages.

People continued to make news. A local resident went south for a trip and when he came back complained that everyone he saw in Southern Ontario looked old. No wonder! Everyone over forty was considered an old-timer in Kirkland Lake in 1926. Charlie Keeley was one of the old-timers who came to live in the Federal area of town. He had found the Keeley Mine in South Lorrain but had no connection with his namesake mine and lived on the gifts of his former partner. They had to be careful in their kindness, for Keeley was fiercely independent and not interested in welfare.

Since 1919 a million dollars had been spent in largely unsuccessful attempts to find consistent ore bodies at Kirkland Lake Gold Mine. By 1924 the shaft stood at 1,600 feet, but values of gold milled were low. Then the company hired a new president and the faltering mine on the top of Beaver Hill began to perk up. Joseph Tyrrell has been called the greatest Canadian land explorer of his day. He had filled in vast hitherto empty areas on the maps of the West as a member of the Geological Survey. In 1883 he found the huge coal seam in the Red Deer Valley and also the great dinosaur skeletons there. With his brother, he crossed the Barren Lands of the North West Territories and became the first white man to do so since Samuel Hearne. In later years he had practiced as a mining consultant, and his reputation and experience attracted investors. He believed that the Teck Hughes ore dipped sharply and would be found in the Kirkland Lake Gold property at depth. Despite continual criticism, he continued shaft-sinking and in 1926 located gold at the 2,475-foot level. From that point the mine was never rich, but it was a steady producer.

A small property that J.B. Tyrell proved. The Kirkland Lake Gold Mine, 1926.   – PA-15526

The Kirkland Rand Mine, 1926. The buildings may have looked neat but the mine was a dud.
– PA-15536

Many of the buildings remain but the Teck Hughes Mine was demolished in the sixties.
– Geological Survey of Canada, 1929, OA-S17943

It is not always easy or necessary to make comparisons among mining camps. Kirkland Lake did not have the large-scale operations that were in place in the Porcupine, and the physical area of the camp was not as big. But on the average the ore bodies in Kirkland Lake were richer than the camp to the north.

For the technical minded, milling in Kirkland Lake was done by sliming cyanidation. The ore was crushed by rod and ball mills, or rolls, and then fine ground in cyanide solution in tube mills in closed circuit with classifiers. Agitating the solution came next, then thickening, counter-current decantation and finally precipitation with zinc dust. The ore did average a high grade of gold, but there were also high tellurides, which meant at first that much gold was lost in the tailings.

The Kirkland Lake mines provided the community with a healthy $1,800,000 payroll in 1926. The Teck Hughes paid a dividend of ten cents a share and the Wright-Hargreaves Mine was now at 1,375 feet. The big news was provided at the west end of the camp. Consulting engineer Robert Bryce of Toronto became interested in the area. He reasoned that if that dipping gold which progressed lower through the Teck Hughes Mine to Tyrrell's newly rejuvenated Kirkland Lake, gold might continue west in the same manner. There was no surface gold and the whole venture was based on theory, which is still an expensive way of naming a gamble. Harry Oakes and his partners retained the two claims he had staked fourteen years before, and with the Elliot claims a company was floated with the magic of Oakes' name as president and the 3,500,000 shares at a dollar each sold briskly. The new mine was called Macassa. One of the puzzles of the camp is that no one has yet come up with an explanation of the origin of that name.

Other developments were taking place at the Lakeshore Mine apart from its rapid growth underground. A substantial fireproof bunkhouse was built across the road from the mine and a large greenhouse in front of it provided fresh vegetables for the residents. The Chateau burned. Old-timers can still recall Mrs. Oakes urging men to throw snowballs on the flames. For Oakes this was no real disaster, as he felt the need for larger accommodations anyway. He projected a nineteen-room residence, and foreman carpenter Dave McChesney was to supervise the construction. Huge logs thirty to forty feet long were used in the work. The building was well on the way when Oakes decided that he needed a billiard room. He solved the problem by ordering that one wall be removed and located further out to give the gentlemen more room to manoeuvre their cues. McChesney protested the removal of the fine logs in vain and they had to go. But at quitting time the building was open to the elements. New furniture was being stored in the yet unfinished second floor. In the night dogs entered through the open wall and workmen were treated to the famous Oakes temper the next morning when he came to inspect the progress. The dogs had ripped the burlap surrounding the furniture and a fine film of eiderdown from the chairs and chesterfields inside was coating everything in sight. Harry fumed and wondered if the dogs belonged to any of the men, but the culprits were not located and the furniture went back to Eatons.

Total work force in the camp was 1,460, while the town had grown to 3,600 persons. The high proportion of miners indicates that single men still outnumbered the married workers. Things were booming at the Teck Hughes Mine with 270 on the payroll. At the Sylvanite a 200-ton mill was opened and when the shaft reached 1,500 feet, steady ore values were encountered. The output that first year in the late-developing mine was $429,424. There were problems at the Macassa. The gold was not in evidence yet and another million and a half shares were floated to bring in more working funds.

A newspaperman, later to become famous in his own profession, "discovered" Kirkland Lake in 1927 and was to be blamed for some of the town's misfortunes in years to come. Gordon Sinclair likened Kirkland Lake to a dairy, with all its white buildings and neat appearance. Old-timers have it that Sinclair did all his research from the verandah of the Ashcan and could not see far enough from that point to notice the white covering was tar paper and not white paint. One of his columns went, "You never saw a busier little town in all your life . . . hardware going into the stores as fast as it can one way and coming out the other, merchants so busy buying from travellers they can't find time to sell; motor trucks and cars, bustling streets and people walking on the mud pavement in the gully; new buildings going up, prospectors and mining men in the half soldier, half hunting garb, in groups everywhere, talking furiously."

Sinclair went on to outline the early days of the gold camp. He compared Kirkland Lake favourably with Cobalt, which he felt looked as if it were recovering from an earthquake. He noted the infectious air of optimism in the town, but the *Northern News* found enough to criticize in the articles. The editor took issue with Sinclair's visit to Timmins and the conclusion that Kirkland Lake was behind the Porcupine in Development. Why Kirkland Lake was embarking on cement sidewalks, had two brokers and two flower shops. One lady had even picked roses in October. This had to beat Timmins, and the town that stands on gold was a model for the "frozen north."

A timely amendment to the Mining Act was hailed as long overdue in 1928. Underground foremen now had to give and receive orders in English. How many accidents might have been prevented if the regulation had been in force earlier. This requirement was soon made mandatory for deckmen, cage tenders and hoistmen. Teck Hughes shares were selling briskly at sixty cents. By contrast, on November 28th the Tough-Oakes-Burnside Mine went bankrupt and closed. The shaft was at 1,900 feet and $2,909,597 had been realized in gold production to this time. There was still gold on the property and it would open again one day.

Loneliness often wore down men in the bush, but just how sincere was this notice found on a prospector's cabin, deserted and the door swinging, its owner long gone? "Fore miles from a nabur; twenty-five miles from a post office; twenty miles from a r.r.; one mile from water; God Bless our home, but I'm glad I'm leavin."

The story sounds of familiar vintage to the one about two miners on Government Road discussing the closing down of a local development property. "No wonder they went broke and had to quit. Why the table they set there would have made the Royal York look like one of them sidewheeler lunch rooms that causes all the greasy sleeves that you see down in Toronto. They called the bullcook a maitre de hotel or something. The grub was wonderful but they had so many foods and tools strewn around a man's plate I had to get my wife to subscribe to *Vogue* for me so I wouldn't offend the boss by using anything but a fork for the ice cream and nothing but the right spoon for the grapefruit. It used to be rather embarrassing for a guy trained to eat off the oilcloth on the kitchen table at home, and the last straw was when the chef demanded the new cager be fired because the poor sucker drank the water out of the finger bowl."

New buildings and activities commanded attention on the landscape. The mines looked like great timber yards as huge piles of wood awaited transport to the depths to provide solid support for new and old workings. Lakeshore even had its own inclined shaft just to accommodate the passage of lumber. But if there were items of interest on all sides, the world beyond the camp was rocking to the disaster of the great stock market crash. Locally the Macassa had trouble just keeping in operation. Investment capital everywhere had slowed to a trickle, and the gold had not shown up yet in economic quantity. By cruel contrast, at the Lakeshore Mine each successive day brought new ore. The big mine kept producing and the mill was now operating at 1,000 tons a day. Close to 300,000 ounces of gold were produced in 1929 at the mine Harry Oakes had found. With 2,140 miners supporting a population of 7,200, the town was flourishing, but a brush with the Depression was just around the corner.

Assay office at the Teck Hughes Mine. Built to withstand fire, it stood the test of time and may be seen today.
– Museum of Northern History

Government Road, main thoroughfare of the rich Kirkland Lake gold camp, 1927.

– Author coln.

MINES OF KIRKLAND LAKE and AREA

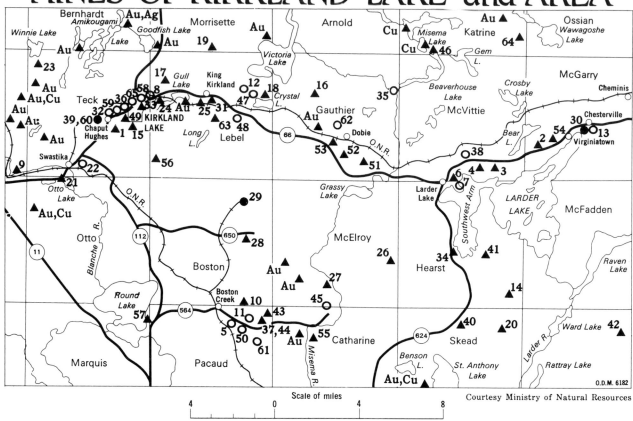

Scale of miles

4   0   4   8

Courtesy Ministry of Natural Resources

O.D.M. 6182

Legend:

▲ Occurrence
● Producer
○ Past Producer

Ag – Silver   Pb – Lead
Au – Gold   U – Uranium
Cu – Copper   Zn – Zinc
Fe – Iron

1. Amalgamated Kirkland Mines . . Au
   Amalgamated Larder Mines
2.   Barber-Larder occurrence . . . Au
3.   Cheminis occurrence . . . . . Au
4.   Fernland occurrence . . . . . Au
5. Amity mine . . . . Cu
   Argosy Mining Corp.
6.   New Laguerre occurrence . . . Au
7.   Raven River mine . . . . . Au
8. Associated Arcadia Nickel Corp.
   (Toburn mine) . . . . Au
9. Baldwin Consolidated Mines . . . Au
10. Bargnesi Mines. . . . . . . Cu, Au
11. Barry-Hollinger mine . . . . . Au
12. Bidcop Mines (Bidgood mine) . . Au
13. Chesterville Mines . . . . . . Au
14. Combined Larder Mines, The . . Au
15. Consolidated Canorama Explora-
    tions (Hudson-Rand) . . . . Au

16. Consolidated Northland Mines . . Au
17. Continental Kirkland Mines . . . Au
18. Crystal Kirkland (Max Kaplan). U, Au
19. Dolsan Mines
    (Mallard Lake). . . . . Pb, Ag, Cu
20. Fabis (LaFond) occurrence . . . Au
    Gateford Mines
21.   Crescent occurrence. . . . . . Au
22. Golden Gate mine . . . . . . Au
23. Gauthier (Winnie Lake) . . . Cu, Zn
24. Glenora (Albert Kokotow). . . . Au
25. Harrison (Kirkrovale) . . . . . Au
26. Hearst-Larder (Detfield
    Lowe and Emil Chorzepa) Zn, Pb, Cu
27. Hennesey occurrence . . . . . Au
28. Jalore Mining Co. . . . . Fe, Pb, Zn
29. Jones and Laughlin Steel Corp.
    (Adams mine) . . . . Fe
30. Kerr Addison Mines . . . . . Au
31. King Kirkland Gold Mines . . . Au
32. Kirkland Minerals Corp. (Kirkland
    Lake gold mine) . . . . . . Au
33. Kirkland Townsite Gold Mines. . Au
34. Korola-Larder Mines . . . . . Au
35. Upper Beaver Mine (Argonaut)
    (Lake Beaverhouse). . . . .Cu, Au
36. Lake Shore Mines . . . . . . Au
37. Lebon Gold Mines . . . . . . Au
38. Lomega (Omega) mine . . . . . Au
39. Macassa Gold Mines . . . . . Au
40. Manor occurrence . . . . . . Au
41. Martin-Bird Gold Mines . . . . Au

42. Mathias occurrence . . . . . . Cu
43. Miller occurrence. . . . . . . Au
44. Miller Independence . . . . . Au
45. Mirado Nickel Mines (Cathroy
    mine) . . . . Au
46. Misema Lake Mining Corp.
    (Forwood). . . . . Au, Cu
47. Moffat-Hall mine. . . . . . . Au
48. Morris-Kirkland Gold Mines. . . Au
49. Northgate Exploration
    (Kirkland-Hudson Bay) . . . . Au
50. Patterson mine . . . . Cu
51. Princeton Gold Mines (Ritoria
    mine) . . . . Au
52. Queenston Gold Mines . . . . . Au
53. Queenston Gold Mines (Anoki
    mine) . . . . Au
54. Rio Algom Mines (Armistice) . . Au
55. Riverton Gold Mines (Gold Hill) . Au
56. Shelp (Dane Copper) . . . . . Cu
57. Sheroomac Mining Corp.
    (Round Lake copper) . . . Cu, Zn
58. Sylvanite gold mine. . . . . . Au
59. Teck Corporation (Teck-Hughes
    mine) . . . . Au
60. Tegren Goldfields. . . . . . . Au
61. Trethewey-Ossian mine
    (Mrs. Claire Cameron) . . . . Cu
62. Upper Canada Mines . . . . . Au
63. Upper Canada Mines
    (Pawnee-Kirkland) . . . . . Au
64. Wadge Mines (Walsh Katrine) . . Cu
65. Wright-Hargreaves Mines . . . . Au

– Ontario Ministry of Natural Resources

204

A Combination of Depression and Boom

My favourite colour, gold!
> – Roy Thompson, who had his start in North Bay,
> Timmins and Kirkland Lake, knew that the yellow
> metal backed many of his enterprises

I was paying eighty per cent of the gold I found in taxes.
Man don't work for that.
> – Harry Oakes hated greedy governments
> as much as speculators.

My object in purchasing the Globe *was not to make money*
. . . I thought I could do something for the country by making
our mining industries better known.
> – Bill Wright

The paradox in mining is that men build headframes that reach high into the sky in order that they may reach deeper into the earth. The thirties saw this trend accelerated in Kirkland Lake and other camps as the price of gold, which had been around $20 an ounce for over a hundred years, started to climb.

The deepest mine in the camp was Kirkland Lake Gold at 4,750 feet. Six hundred tons a day were milled at the Wright-Hargreaves and the mill was enlarged. The Teck Hughes produced 1,000 tons a day for milling. Its 446-acre property had the lowest costs of production of any gold mine in Canada at $8.305 per ounce. In 1930 the new Teck Hughes shaft had cost $114 a foot to sink, in an excavation thirteen by twelve and a half feet. The patient years of the Sylvanite shareholders paid off as the mine announced its first dividend of four cents a share. At that property a series of faults had dislocated the upper levels and it had taken years and one and a quarter million dollars to prove the mine. Lakeshore was the biggest of the mines and would remain so. It milled between 1,500 and 2,300 tons a day. Huge quantities of sand from Crystal Lake were used to backfill the mine. Seventeen million dollars worth of gold was extracted in 1930, and it provided $6 million in dividends. At the same time the property at the east of town was going into its third life. The Tough-Oakes-Burnside was now reopened as the Toburn, an amalgamation of the initials of the other three names. The next year the dewatered property milled 125 tons a day. Its elegant headframe still graces the skyline.

Harry Oakes was busying himself with his half-million-dollar home in Niagara Falls. He would visit Kirkland Lake with his consulting engineer, Montague Barney, but could credit no man but himself with his success and said he should have called his great mine Oakes Consolidated. Those five Lakeshore claims alternated almost yearly with the Hollinger as the largest gold mine on the continent for some time to come. In 1931 the combined output of the Kirkland Lake mines exceeded that of the Porcupine, and the Lakeshore milled 533,757 ounces. Total production for all gold mines that year in the Northeast was $8 million and this was in the height of the Depression. All three of the camps explored in this book produced $200 million alone in dividends between 1904 and 1931.

The Lakeshore was producing $36,000 a day or $25 a minute, and Oakes decided to unload his Macassa shares as perhaps having little promise. He had resigned as president and received $50,000 in shares for his interest. These were sold for a dollar each, but he was to bitterly regret the move when they rose later to $8.50. Robert Bruce took over as controlling executive and the work of proving that mine went on. Over at the Teck Hughes, the shareholders were wondering about sinking the shaft to depth, but Doctor Forbes, the manager, reassured them. The mill was now operating at 1,300 tons per day and the mine was at its peak at $6,093,199 in gold produced in a year.

In contrast with the fortunes being produced from the ground of this once painfully slow-moving camp, the relief rolls had climbed and men lived in hobo jungles around the community. Mine cyanide cans, scrap lumber and nails provided their makeshift homes. Residents said harsh things about Gordon Sinclair for his articles, which they said encouraged the drifters to come to the gold towns. He was right in his estimate of the riches the camp would provide when it came into its own. In the midst of the Depression gloom the combined Porcupine and Kirkland Lake output average $1 million worth of bullion every ten days. By 1932 Kirkland Lake had produced a total of $100 million in gold and the annual payroll was $4 million.

The *National Geographic* had kind words for the North. The writer likened the bush country to a rocky crust with thin dirt on it and a myriad of puddles growing into lakes and billions of trees.

"You think of Bret Harte's *Roaring Camp* when you see Kirkland Lake on a Saturday night. In crowded, crooked streets a dozen men to every woman, stores open till midnight, even the hardware and furniture stores. Some sacrifice to Bacchus but very few idle men. Finns and Chinese wearing twenty dollar gold pieces as watch charms; a crowded movie showing *Two Nights in a Barroom*; brawny Russian miners sprawled in barber chairs getting an over Sunday polish; the smell of fresh cut pine and the noise of saw and hammers as bohunks work by floodlight on a new hotel; young engineers in caps, sweaters and high laced boots socially playing cards in a crowded lobby, slapping cards down noisily." The breathless emissary of the world's largest magazine was astounded that no one would bring his bags into the hotel — the Ashcan — and he had to carry them up two flights of stairs. But the excitement of the gold camp did not prevent recording of the fact that one of the mines had produced a record $122,000 in gold that day.

With activity came contrast. General Manager Maurice Summerhayes of the Wright-Hargreaves Mine was forced to place a large advertisement in the local paper. It had come to his attention that attempts were being made to purchase employment, and this was not possible. There may not have been many idle men on the main streets, but those who were out of work haunted the mine gates. Bill Wright had seen hard times, but for him life was now full. He was called soldier, prospector, pioneer and gentleman in news articles. In 1932 he was enjoying a large breeding farm and racing stable, and his horse had won the King's Plate.

Bob Bryce's patient search paid off. The Kirkland Lake Gold Mine had obligingly put forward its 2,475-foot level into the Macassa ground and ore was found. The Macassa number one shaft intersected this area, and by 1933 the latecomer was milling gold. Over at the Wright-Hargreaves, the mill was handling 1,000 tons a day. Neighbours to the Macassa, the Kirkland Lake Gold shareholders were also patient people. Their long wait of nineteen years was satisfied in 1934 when the mine paid its first dividend. The thrice-born Toburn mine settled in to an uneventful nineteen years of productive life.

The first head frame at Macassa Mine nearing completion in 1932. The mine is still an active producer.   – PA-14699

This 1934 view shows how close the Wright Hargreaves Mine was to Government Road, Kirkland Lake's main street.   – Airmaps Ltd., PA-15689

Chief Dennis Tuohey of the Wright-Hargreaves security department charged an employee with highgrading. The man had taken rich ore from the 3,300-foot level of the mine. A search of his room revealed another two pounds of high-grade. While the gold fancier awaited trial, the town police did the only natural thing in a gold camp. They sent the stolen ore away for assay so that the total value of the gold could be clearly sworn at the trial.

By 1935 the mines of Kirkland Lake had produced $163,000,000 in gold, paid $60 million in dividends, and their total market value was $250 million. With gold at a record $35 an ounce, 3,490 men could be employed in the mines. Visitors to the town at night saw a bright reflection in the sky from the lights of seven mills that never stopped. The Lakeshore milled 2,000 tons a day and was one of the biggest gold mines in the world. Shares that once went for forty cents commanded $64 and gave dividends of $6. Figures like 12 million board feet of lumber sent underground for timbering tell the story of development that year. The Sylvanite mill averaged 432 tons a day using new ore found close to the Wright-Hargreaves boundary and the mine paid dividends of twenty cents s share. The Wright-Hargreaves Mine was going full tilt with twenty-eight operating levels. At Macassa a grateful board of directors voted Bob Bryce 50,000 shares at a value of one dollar each for his work in establishing the mine. Only at the Teck Hughes was there a hint of gloom. Production slowed a bit and the mill dropped to a thousand-ton output.

Harry Oakes' Chateau in 1936 was used solely for Lakeshore directors' meetings and entertaining.   – PA-17717

Gertrude Oakes, Harry's sister, was lost at sea in 1935. She was drowned when the Ward liner *Mohawk* sank off the New Jersey coast. Gertrude had never married. Once the great Lakeshore Mine was established, she had moved into the area and become closely involved with its affairs as director. She had lived at her Red Pines log house at nearby Kenogami and had just finished arranging Christmas gifts for mine employees before setting out on the ill-fated cruise. Gertrude Oakes was buried in the family plot in Maine and was joined by her brother only eight years later.

An examination of the Lakeshore Mine report to shareholders for 1935 makes no reference to Harry Oakes. The mine owner had resigned as president and was off to London to be presented at the Royal court, twenty-three years after he had pitched his tent on the shores of Kirkland Lake as a penniless prospector. Harry had what he felt was a legitimate beef with the Canadian government by this time. He had given land in Niagara Falls to the people and had been assessed $250,000 in back taxes. He reckoned that he would end up paying $3 million a year in income tax. Oakes spoke in his usual blunt manner. "Pride of ownership used to belong to all men, but it's getting narrower. Pride of possession belongs to the politicians. You find it, they take it." Chatting one day in Swastika with George Ginn, the mining recorder, he stated that the real reason he would have to leave Canada was the succession duties. If he died, his stock would be forced down on the market and his family would be left much less secure than was the present case. He announced the departure of the Oakes family to take up residence in the Bahamas, where he could find relief not only from excessive taxation but also from chronic bronchial trouble. The newspapers echoed a welling up of public opinion critical to his attitude, but Harry bluntly had the last word. "I found the pot of gold at the end of the rainbow and I found it in Canada. But I was paying eighty per cent of the gold I found in taxes. Man don't work for that."

Roza Brown's house, 1934, with the Wright Hargreaves Mine in the back right.
– Museum of Northern History

Bill Wright was living in Barrie, still racing horses, and his latest philanthropy was a gift of $200,000 to the Barrie hospital. George McCullough interested him in buying into the *Globe and Mail*. He did so, but only as a silent partner. As usual his action reflected considerable thought. "My object in purchasing the *Globe* was not to make money out of it, at least that was not my main object. I thought I could do something for the country by making our mining industries better known. Anything that is of advantage to mining is of advantage to the country as a whole." Reporters sought him out and asked his opinion of Oakes leaving Canada. Bill Wright gave them no satisfaction. He said he felt Harry's action was justified, but he preferred to stay in his adopted country.

Roza Brown always felt her downfall occurred in 1937, in her eighty-third year. Curiously enough it all happened over a straightforward real estate deal. She had bought many lots, but none had appreciated so much as the rocky corner property on Government Road which had originally sold for $100. The tough old girl repeatedly refused to sell, but when she was offered $30,000, she finally accepted the deal. The sale was to haunt her. When she saw a huge basement going up on the spot and learned the S.S. Kresge chain was building, Roza's wrath knew no bounds. She sued all and sundry who had any part in the transaction. But it was a fair deal and she had been outsmarted. She even went to see magistrate Atkinson in his chambers, and when he refused to debate the issue, called him several unprintable names and left for home. When a Toronto bailiff attempted to serve an order forcing her to surrender the deed, she threw a shovel at that worthy and he beat a hasty retreat.

The 4,640 mine employees in the gold camp produced record ore and a handsome $20,260,000 in dividends in 1936. This was the peak production year for Kirkland Lake. The Macassa rejoiced in having recovered $2,602,226 in gold by 1936. The Wright-Hargreaves Mine had so far paid out over $42 million in dividends and a new shaft ensured a good supply of ore for an extended period. The Lakeshore Mine still led the pack, and by the fall of 1936 just over $99 million worth of gold had been recovered. The total dividends of $53,020,000 was only exceeded by that of the Hollinger.

Kerr Addison Gold Mines in Virginiatown had gone through several phases before a satisfactory ore body was outlined. In 1938 the great mine, in which the original stakers now had no interest, was well on the way to heavy production. Back in Kirkland Lake, the huge 220-foot smokestack of the Lakeshore Mine went into operation. One hundred and fifty miles south, George Minaker was killed by a runaway horse while skidding logs near North Bay. An old picture of Minaker survives, too time-worn for reproduction. The print shows a slim, clean-shaven, forthright-looking man. He was all of that, but he never really made big money from mining. After the sale of the claim that became an important part of the Lakeshore Mine, he went back to the woods. His one other claim to fame was when he drove Haileybury founder C. C. Farr and his wife to a meeting in a cutter and the vehicle overturned, depositing them in the snow. He made money in a backhanded way when a syndicate took over his remaining claims and sold them to the mine for the south slimes holding basin, but by then he had lost most of his funds in the 1929 crash. He never bemoaned the sale of claim L.16653 to Oakes. To him it was a deal and best forgotten.

In 1939 the two founding pioneers renewed their acquaintance with the town. W.H. Wright donated $100,000 for the Wright wing at the hospital. It meant fifty more beds, and an editorial in the *Northern Daily News* commented that the gift was unsolicited and therefore all the more appreciated. Harry Oakes arrived in the person of a large, framed portrait. In recognition of philanthropies in England, he had received a baronetcy, and in typical Oakes fashion sent the colour portrait in full regalia to the Lakeshore Mine. The picture is Harry, triumphant, jaw stuck out and challenging all comers. As mine officials pondered what to do with the unexpected gift, a deed for the lot occupied by St. Peter's church arrived by mail. Harry never visited Kirkland Lake again.

GOLD BULLION
KERR-ADDISON GOLD MINE

– E. Duke, Author coln.

KERR-ADDISON GOLD MINE
VIRGINIATOWN, ONT.

The great Kerr Addison Mine at Virginiatown came into its own in the late thirties.
– E. Duke, Museum of Northern History

When the King and Queen made their tour of Canada, Roza Brown went to Montreal to see them land. She had no invitation, but a piece of paper was of little significance to the diminutive Hungarian-born pioneer. Roza was a strong royalist and not only frequently sent telegrams of birthday greetings, but usually carried framed portraits of the royal family. She arrived at the docks replete with a dress reminiscent of the turn-of-the-century style which sported a Union Jack as a sash. She carried a couple of other small flags and in honour of the occasion had forsaken her usual rubber boots in favour of more elegant footwear. No one knows how she made her way through the guards and located a prime position near the main gangway of the ship. A retired commissioner of the R.C.M.P. has written that he checked the little old lady out, decided she was harmless and escorted her to a seat. But when people in front rose to see the royal couple better, the resourceful northerner revealed her secret weapon. The ends of her little flag sticks had been sharpened to cruel points. With aim made more accurate by her feeling in the matter, Roza struck the posteriors of those obstructing her special view. By the time the royals swept by, their loyal subject could make her practiced curtsey with no one either in front or to the side of her.

Kirkland Lake was a strong sporting town. The mines had sponsored both athletes and semi-professional teams. Foster Hewitt referred to Kirkland Lake as "the town that made the N.H.L." Roy Conacher, younger brother of famed Charlie Conacher, left the Wright-Hargreaves organization to play for Boston. The parade of hockey stars from the town continues to the present day. Millionaire sports promoter Jack Kent Cooke rounded out a term as manager of Roy Thompson's radio station, Kirkland Lake's CJKL. Just before the war, the Lakeshore Blue Devils won the coveted Allen Cup.

The year 1939 was a great one for the town. The population peaked at 24,200 and at 4,680 there was the greatest number of mine employees ever. The town would never be as rich or as busy again.

Hard work rewarded. Sir Harry and Lady Eunice Oakes.
– Museum of Northern History

These Lakeshore Mine houses look as good today as they did in this 1936 view. — PA-17716

Looking west towards Beaver Hill along the mile of gold. — E. Duke, Author coln.

After a shower, miners could get a tan in the solarium at the Wright Hargreaves Mine, 1936. – PA-17579

Compressor house at the Lakeshore Mine.   – Museum of Northern History

Sand dumped into the chute is distributed over the whole mine as back fill for stopes. Lakeshore Mine, 1936.
– PA-17616

Slime settlers in the thickener tank at the Lakeshore mill.
– Museum of Northern History

Underground
Kirkland Lake Ont.

– E. Duke, Author coln.

After the Peak Years

*The acquisition of riches has been to many, not the end of
their miseries but a change in them; the fault is not in the
riches, but in the disposition.*
> – Seneca, the Roman philosopher, did not have
> Harry Oakes in mind, but the idea is about right.

*Though wisdom cannot be gotten for gold
Still less can be gotten without it.*
> – Samuel Butler

In 1940 the federal government needed gold to pay for war material produced in the United States and both the Porcupine and Kirkland Lake camps were spurred to higher production. Total gold produced in Canada that year came to 3,200,000 ounces, of which one quarter was from Kirkland Lake. Throughout the early war years, the Kirkland Lake theatres put on "Stamp Out Hitler" shows. Admission was by war savings stamps. Roza Brown took pleasure in these events, as she purchased savings bonds, and went up on the stage and kicked Hitler's photograph. Another person who had more than his picture assaulted was T.B. McQuestin, Ontario Minister of Highways. He attempted to get the name Swastika changed to Winston in an obviously patriotic gesture. The small community clung to its name. Local reasoning was that the village had been named longer than the dictator had owned claim to the crooked cross. New signs were torn down as fast as they appeared, public meetings protested the move and poor McQuestin supplanted Hitler, at least in Swastika, as public enemy number one. Officialdom conceded defeat, Swastika remained and the embattled minister was pleased to slip into oblivion.

Even in the war years, Kirkland Lake made mining and labour news. The Wright-Hargreaves Mine developed pneumatic guides to allow smoother travel for skips carrying ore to surface. Tiny tires made the movement smooth and noise free. But it was not for innovation that Kirkland Lake was in the news in 1940. The first strike in the community in 1919 was quiet and uneventful. Not so that which took place twenty-one years later. It had a start at the Teck Hughes and spread to the other mines. The issue was wages and working conditions, but the timing was not right. From the 1939 peak of nearly 5,000 miners, more than 1,700 enlisted when war broke out. Gold markets were on the edge of collapse anyway during the war and the provincial government saw the strike as a direct challenge to its rule. The local thirteen-man police force handled the situation but was overruled when 183 provincial police officers were sent to the community to keep the peace. Between 8,000 and 9,000 people moved away during the three-month-long work stoppage or did so in the next few years. Gold mining was not an essential war industry and the strike neither stopped production nor prevented the inevitable decline of ore. The miners had the support of the Canadian labour movement, but it was ineffective. Timmins miners sent a cheque for $1,000 to aid the strike fund. Once during the strike there was a rock burst at the Lakeshore and miners left the union hall to help rescue trapped men when even by their terms the men working were scabs. The strike ended in February 1942 with approximately a thousand less miners in the camp.

Once gold was reclassified as a non-essential war industry, production fell off quickly by eighteen per cent. A premium was offered to boost the price of gold to $38.50, but there were only 2,063 miners working in Kirkland Lake by the time peace came. The mines were able to increase production as area men returned from the war. The Wright-Hargreaves began using a minimum of 100,000 board feet a year to secure its workings. The Lakeshore mill worked up to 1,000 tons a day again. At Kirkland Lake Gold, $25,000 was invested in new dams for slime dumping. Gold-camp life seemed back to normal.

Sandy McIntyre never did make his third big strike. He died in 1943, a shadow of the man who once packed into the Porcupine. Another pioneer died in 1943, but his end was not peaceful. The sixty-seven-year-old Harry Oakes was murdered at his palatial home in the Bahamas. The killer burned the body and sprinkled chicken feathers in a grotesque imitation of voodoo rites. There was a messy trial of his playboy son-in-law, but the case fell through and the real murderer was never caught. There are many theories about why Oakes was killed, for no real solid facts about the actual killer emerged. The most plausible theory so far is that he was put to death by Mafia-related interests. Harry had blocked attempts by organized crime to set up gambling on the islands. After his death, the place became wide open for casinos. Oakes had done much for the Bahamas, but his blunt pugnacious air made him as many enemies there as it had elsewhere. His estate left $14 million, much property and a nagging feeling that there must have been more money somewhere.

Harry Oakes had been able to provoke heated discussion during his life and even today he is able to provide grounds for discussion and argument. He was a man of considerable erudition, which was cloaked in a boorish facade of the self-made man. He had paid heavily for his hard years of grubbing for money and remained until his death essentially suspicious of his fellow man. He gave three Spitfires to aid the war effort and his name appears on a highway and an airport in the Bahamian capital, yet he is not fondly remembered. His home there is now gone and a Casino is situated within the Ambassador Beach Hotel on the former grounds. One wonders if the killer lived long enough to play in that place. But consider the man and his legacy. He founded one of the greatest mines ever known on this continent, proved its worth, brought the property into operation and controlled it for years. He is said to have made the largest single fortune ever garnered by an individual from Canadian mining. His endeavours contributed in a major way to the establishment of a town that lives on. Remember him, warts and all, as a man who made a contribution to Canada.

Roza Brown passed away in 1947 at the age of ninety-four. She left provision for her ever present dogs and had a great funeral, which would have given the old reprobate much pleasure. Roza Brown was a liberated woman before the term was ever invented. The mining community where she spent half her life kept right on working.

Two men who had left Kirkland Lake long before, but still maintained an interest in it, passed away soon after. Ed Hargreaves had become his brother-in-law's secretary and for several years his family had lived with Bill Wright at 55 Peel Street in Barrie. He died at age seventy-four in 1950, and the following year Wright passed away at the age of seventy-five. Bill Wright had financially assisted both St. Peter's Church and the Royal Canadian Legion. The slim, balding man with well-clipped mustache had retained his unassuming manner to the end. Wright kept his prospector's kit packed and ready to go in the bedroom of his Barrie mansion. If his wealth might one day desert him, he was ready to hit the road again. The discoverer of a great mine and co-founder of Canada's national newspaper died well respected and without any enemies in sight.

Mine finder, Dr. Bob Bryce is back left in this group at the Macassa Mine. — Museum of Northern History

William Wright was said to be the richest private in World War I. — Museum of Northern History

Kirkland Lake's eccentric Roza Brown admired Royalty – and Harry Oakes. — Teck Centennial Library

Kirkland Lake continued to produce gold in good quantity, but the camp was inevitably slowing down. In 1953 Kirkland Lake Gold Mine paid its last dividend and the Toburn closed. Its total production had been a very respectable $17,738,506. Al Wende ended his thirty-nine-year association with the Lakeshore Mine when he resigned his directorship in 1956. That was the time Little Long Lac Mines took over the property and others in the area.

J.B. Tyrrell, who still visited his former mine and went underground in a wheelchair, died at ninety-nine. The man who had crossed the prairies part of the way in a covered wagon seventy years before was given the epitaph of Canada's senior geologist.

Another man with a passing acquaintance with the Kirkland Lake area through Boston Creek visits died violently in 1957. Albert Anastasia was gunned down by Crazy Joe Gallo in the barbershop of the Park Sheraton Hotel, New York, a victim of his Mafia-related connections.

Al Bargnesi, third from the left, founded a one man mine at Boston Creek. Albert Anastasia, on his right, was his brother-in-law, murdered in an underworld crime dispute.
– Desmond Woods

Round that year out with a chuckle. The Royal Bank was put on the spot in the winter when a train carrying the Lakeshore Mine payroll was held up north of North Bay by a blizzard. The bank manager mentioned to Charlie Chow that the necessary funds were not available to fulfill mine obligations. Within minutes the hotel owner provided the entire sum, all in bills carried in brown paper bags. Over the next few years attempts were made to rob Charlie's Hotel, but no money of any consequence was ever taken.

It is never pleasant to dwell on decline. The Lakeshore was milling only 480 tons a day in 1960. Both it and neighbouring Wright-Hargreaves were administered by Little Long Lac Mines. Kirkland Lake Gold closed. Since 1919 it had produced $39,124,929, not bad for a mine which had taken so long to establish. The Sylvanite followed in 1961. It had given up a total of $56,596,502 in its working life.

Macassa pioneer Bob Bryce died in 1963 at the age of eighty-two. He would have been pleased with the progress of the mine which he had patiently shepherded into production. The Macassa operates today and has expanded far beyond its original property.

The Wright-Hargreaves closed in 1965 and the Lakeshore followed shortly after.

The mine Bill Wright had founded closed with a total of $157,308,926 produced from its claims. The Lakeshore milled Wright-Hargreaves ore at the end, and when its doors closed, the great mine Harry Oakes put together had given up a staggering $265,000,000 in gold. The third greatest of the camp mines, the Teck Hughes, closed in 1968 with a final total yield of $105,320,778. In the last eight years of operation it had controlled the Kirkland Lake Gold Mine.

The Wright-Hargreaves wound up being the deepest gold mine in North America at 8,172 feet, with the Lakeshore only twenty feet behind. As the mines grew older, rock bursts and cave-ins became common. Even some with ore still blocked out became uneconomical to work due to lengthy tramming and as much as three hoisting sessions to bring ore to surface. Even in these days, when such figures tend to numb the senses, consider the record of the Kirkland Lake gold mines. A total of $735,697,520 in bullion was produced up to 1969, with one third of that originating at the Lakeshore Mine. Dividends represented more than a quarter of the gross figure. Put those amounts in their correct perspective according to inflation, the dollar value and other market forces, and a vast contribution to the growth of Canada is seen from a small mining town.

The mines that made Kirkland Lake great live on in other areas. The Kirkland Lake Gold headframe — it had a name change from "gold" to "minerals" in the last years — was sold and now does duty for a mine in northern Quebec, as does the former Sylvanite headframe. Equipment from Kirkland Lake mines was salvaged and sold all over the continent. The big aerial tramway across the highway at the Teck Hughes was dismantled in 1969, and a month later the Lakeshore smokestack was blasted down.

Charlie Chow died in 1972 at the age of eighty-six. The popular Chinese, who found his pot of gold at the end of a frying pan, left half a million dollars and much property. Typically for the cautious Charlie, only $400 in cash was found in his room.

Read *The Financial Post* and Kirkland Lake is always cropping up. In an article about an exploration company,. we find that the founder was August Mitto, otherwise known as the Russian Kid. In 1985, in one of the frequent mergers seen in the mining industry, three established companies turned into a new one. The Lakeshore mines, Wright-Hargreaves mines and Little Long Lac mines were reborn as LAC Minerals. LAC is still very active in Kirkland Lake, controlling the Macassa Mine, a new shaft of which will be North America's longest single lift, and the reborn Lakeshore Mine, complete with new headframe and innovative methods of recovering ore from the long-dormant giant mine.

Recall the Tough family that started it all along with Harry Oakes when they staked the Tough-Oakes Mine. George Tough's namesake grandson became the Deputy Minister of Mines for Ontario.

People do not forget J.B. Tyrrell. A $30 million museum opened in 1985 in the Alberta badlands near Drumheller, close to the great dinosaur discoveries of 1884. The building honours the man who found the bones, Joseph Burr Tyrrell.

Today, not only is Kirkland Lake a mining town, it exports mining technology and expertise all over the world. There may be few headframes dotting the skyline, but earlier prophets of doom have been confounded, the town has stabilized at about half its peak population and it is a permanent part of the Northern Ontario scene. The streets are still switchback in places and rock is ever present in the community, but towns like Kirkland Lake survive through spirit, a northern commodity which is never mined out.

Gold sat on the Swastika platform waiting for shipment south. These bars are worth more than one million dollars at today's prices. Bill Brennan, station agent, unknown, Jack Gale, veteran prospector, 1921 or 1922.   – Teck Centennial Library

CONCLUSION

If there is any thesis to *Fortunes In The Ground*, it is simply that the three great mining camps that provided Canada with an abundance of silver and gold begat each other. Cobalt men and money developed much of the Porcupine. Kirkland Lake was strongly linked with the other two camps. All three aided immeasurably in the development of Canadian mining.

Many of the familiar mining firms, if now no longer in their original birthplace, are still on the mining scene. Take Cobalt's Coniagas Mining Company. The outfit is long gone from the silver town but is active in many areas, referring to itself as the oldest continually listed mining corporation in Canada. The Teck Corporation is a giant in the business. Spend some time glancing through *The Northern Miner* or *The Canadian Mines Handbook* and you will find other examples of mines from the three camps visited in this book which are very much alive in the mining industry today.

Mining towns were called camps because no one thought they would last. The doom and gloom merchants that predicted the demise of Cobalt, Timmins and Kirkland Lake have been confounded by these communities. These mining towns diversified into other fields of endeavour, but all three remain mining centres, hosting mines, human expertise and specialized trades and equipment for the industry.

There are ever-new prospects on both near and more distant horizons. The Casa Beradi area of Northwestern Quebec and the Harker-Holloway district east of Kirkland Lake are just two places which will funnel business into the established mining towns. Meanwhile the prospectors look at the ground in between the three camps which are the subject of this work. They reason that the next big strike, or at least a new mine, will have to be in the vicinity of the present camps. Gold is especially difficult to locate. It could be anywhere. Prospectors Don McKinnon and John Larche started the great Hemlo gold rush. The area had been crisscrossed by scores of people over the years, yet they were the ones to find it.

I will conclude with a recent statement by Don McKinnon. He knows that Northern Ontario has a lot more to offer, that the surface has only just been scratched in mining terms.

"I'm just a bushman. And if there's one thing I want to do, it's save the art of prospecting . . . The old-time prospectors develop another sense — the nose or feeling, call it what you want — and if you combine that with all this new technology, then you're going to develop people who can't be beat."

Prospecting and Claim Staking

A good promoter is one who must believe in what he is doing. He must have absolute faith.

— Murray Pezim, the prime mover behind the Hemlo discoveries.

We lead.   — Motto of the Prospectors and Developers Association

Prospectors are ever ready to laugh at themselves. They have to have this facility, for theirs is a rough and often lonely business. Here is what prospectors and others have said about the life in the bush.

Shakespeare may not have thought of prospectors when he wrote these words from *As You Like It*, but they have a ring of truth:

Are not these woods
More free from peril than the envious court?
And this our life exempt from public haunt
Finds tongues in trees, books in the running brooks?
Sermons in stones and good in everything.

Rabindranath Tagore finds a connection for us between the obsession of the man of industry and that of the prospector:

I thought that my invincible power would hold the world captive, leaving me in a freedom undisturbed.
Thus night and day I worked at the chain with fires and cruel strokes.
When at last the work was done, I found that it held me in its grip.

Fred Thompson, himself a prospector, said:

All you need to be a prospector is a strong back and a weak head. One prospect in a thousand makes a mine and even prospects are hard to find.

When success loomed, prospector Lloyd Otto summed it up:

For what does the mirror say of you?
Too old for women and song and wine,
All ready to die but you've found a mine.

Joe Parkin spoke of his fellows in the first decades of this century: "They were independent and did not work by a whistle. A prospector is a dreamer. He could go into virgin bush and know that if he came up with something that is feasible and becomes a mine, he and his partner would be responsible for jobs, a town. Many prospectors just achieved an existence, but still they went out looking for that rainbow. After all, in those days $300 would do as a grubstake for a summer. There were no planes, no outboard motors. It would take a week to ten days to get in where now you can drive in an hour. You'd have to live largely off the land, maybe pick an area, build a cabin and work there for five years or more.

The Australians have an apt word for prospectors. "Fossickers" comes from the Latin *fossa*, a ditch, and *fodere*, to dig. Put that together and you get one who rummages among the rocks in search of ore. Call them prospectors, those who worked from dawn to dusk in bitter cold or scorching heat, through fast rapids and mosquito-infested swamps, their only equipment an axe, hammer, compass and the eternal hunch so peculiar to the North American style of dogged determination. They were jacks-of-all-trades and specialized in optimism. They did not think their lives anything special. They put up with flies, rain, sleet and snow, often poor food, and little of it, worn-out clothes and boots, infrequent changes of underwear, poor equipment, bad teeth, lack of medical care and few creature comforts. Today we find it difficult to understand the drive that led men in search of gold. Try to think of it as the finding rather than what came after. The true searcher kept on going even when the golden rainbow was within his grasp.

The understanding becomes more difficult when prospectors talk about the feel for the country. Gold is found by looking, they say. Walk over the ground from every angle, at every hour, and get that feel for it — even the stones themselves seem to tell a story after a while. One prospector I know keeps rocks of all descriptions in every room in his house. He says it keeps him in touch with the land. The true prospector only does other things to support himself for just another stake into the bush country. Jobs for big exploration outfits are just stopgaps until the chance comes to go out again. Prospectors have their own theories, often untrammelled by book learning, and they sometimes express themselves in obscure ways. One said to me, "I am going to take my packsack and axe this spring and go out to a place I know where the granite has had a fight with the greenstone."

No matter what their mining background, prospectors all agree that the best place to find minerals is in the bush. Murray Watts, one of Canada's greatest prospectors, put it this way: "One of the difficult things with the academic approach to mining is that if you spend all your time working out the mathematics of finding anything, you'll never start looking." Prospectors have to have a very open mind. Robert Flaherty says, "There is a saying among prospectors, 'Go out looking for one thing and that is all you'll ever find.'"

Prospectors do use every aid and offer of federal or provincial government assistance open to them. They use all manner of mining surveys, study the latest on glacial geology to see what the great ice streams stirred up, and make use of what technology they can afford. Geophysics is a great finder's tool today. Magnetometers and electromagnetic instruments provide clues as to the possible locations of deposits. Anyone that has the funds to do the same work by air can obtain ever better results.

The Prospectors and Developers Association and the Canadian Institute of Mining and Metallurgy bring all sections of the industry together. Meetings of both groups are seldom stuffy, for this is an essentially practical business. Prospectors are thinking all the time. Recent discussions centre around whether prospecting is on its deathbed to the reasons why a lone prospector finds it difficult to get a grubstake. That is nothing new, for it has been a problem ever since prospectors went out into the bush. If you want to get into a good argument, suggest the familiar mine manager's axiom that mines are made not found. These are not the sentiments of prospectors. They will tell you that the three camps discussed in this book were found by prospectors not geologists. With a hint of optimism, the commodity which fuels the prospecting life, John McAdam said, "I don't worry about running out of resources. When supplies get low, prices go up, which stimulates production for minerals that are made more economic by the rise in prices . . . gold is good and I think it will stay that way."

Let's leave this section on prospectors by considering what Robert Service said of them in *The Men Who Don't Fit In:*

> *There's a race of men who don't fit in*
> *A race that can't stay still;*
> *So they break the hearts of kith and kin,*
> *And they roam the world at will;*
> *They range the field and they rove the flood;*
> *And climb the mountain's crest;*
> *Their's is the curse of the gypsy's blood,*
> *And they don't know how to rest.*

Maybe the stories of some of the mine finders in this book have stimulated the urge to explore and even stake a claim. If this is the case, make your way to the nearest mining recorder's office. Pick up area maps and fill out an application form for a licence. The sum is nominal and there are no barriers as long as you are over eighteen years of age. Have a chat with the recorder about the legal requirements of claim staking and pick up a summary of the requirements under the Mining Act.

So go to it. There are a lot of precious and base metals out there. All you have to do is find them.

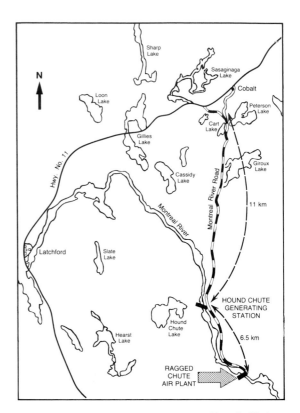

– Ontario Hydro

*The compressed air blow off on the Montreal river,
a unique harness of nature's power.* – OA-S17630

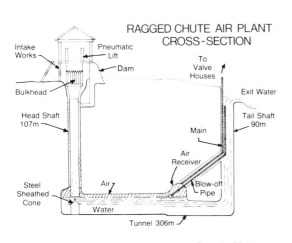

RAGGED CHUTE AIR PLANT
CROSS-SECTION

– Ontario Hydro

The force of the Ragged Chutes, site of the famous compressed air plant, as seen today. – Author

the job, and there's moving the equipment to new sites, and even the unexpected must be considered. The novice is reminded that the industry is very competitive and accurate pricing is essential. He laughs at the complexity of it all. "When I am working things out, the field supervisors come in and tell me a better way to do the job!" Before leaving, he points to drill rig costs. Depending on size and capability, they can run from $65,000 to $125,000. Take a medium-size rig. For $100,000 the firm has in its inventory the rig and supplies, 2,000 feet of rod, 400 feet of assorted casing, 2,000 feet of hose and some core barrels. Add on another $85,000 for a tractor and one is easily convinced that this is a big capital operation. There is the burned out wreck of one such rig in the yard. A freak nighttime accident did the damage and no one was hurt. It is one of the hazards of the business.

A company truck takes the visitor out to a drill site where men are working for an unspecified company on an unmentioned lake some distance from town. The firm's reputation relies on its ability to keep such information confidential. Exploration depends on security for promotion; claim staking and other factors can all be influenced by knowledge kept in the right sphere. The current source of this interest is set up on a big ice-covered lake. Three rigs are strung out on the perimeter. In the middle of the lake a couple of pumps are supplying water for the drills. They are sheltered under portable iron-framed covers. Everything about a drill company depends on portability. The water is pushed through coil heaters to keep it from freezing. Depending on the type of fuel used, one such heater can run $60 to $90 a day, explains the necessarily cost-conscious company man. The hoses fan out to the rigs and are covered with snow for insulation.

As we bounce along the rutted ice road, the guide explains the principle of reverse circulation drilling, a company specialty. This is not what we shall see today, but it is a service which helps geologists to sample swamps and overburden. When the glaciers ground down the earth, they left minute particles of minerals in the upper soil and debris covering the bedrock. Such shallow drilling is relatively inexpensive compared to diamond drill work. The drill can cut through the muck and send up chip samples in a slurry of

This slurry or rock chips and water from a reverse circulation drill will be analysed for mineral content.
– R. Regimbal

Drill cores split in their boxes.   – R. Regimbal

Diamond drill working in rough terrain.

– Heath & Sherwood

water. This can be analyzed and then a determination made both on the content of the overburden and also whether further deep drilling is warranted. Not all drill methods are as complex as this. Drillers can go through shaky ground or ever shifting sands using bentonite, a natural mud substance, or artificial polymers which firm up the ground. The latter agents are used after some consideration, as a bucket full can cost over $200.

Visitors in hard hats drive up to one of the rigs. Imagine a big plywood cube mounted on skids with tripod overhead. The driller and his helper wave a greeting, but they do not hear much, as they both wear ear defenders from the roar of the machinery. The object of all this activity is the drill rods, which run through the floor at an angle calculated to reach the required formation. To one side is the power source, a big GM-motored diesel. In one corner a Bean pressure pump forces water through the system. There is a drum of wire line and to one side a small oil stove. Rick mentions that Heath & Sherwood makes some custom equipment for the industry. This small stove is one example, being extremely popular, and requests for it come from drill firms all over the world.

The water is necessary, for a drill bit will burn up very quickly if it runs dry. There is also a great deal of oil and grease to keep the machinery well lubricated. The drillers do not stand still for long. They run machines, take core out, hoist rods, maintain the fuel supply, read gauges, and do an endless variety of jobs. They must be good take-down and put-up artists, for that smoothly running operation did not get there on its own. Over the din of the drill I asked Mike, the driller, if he got cold. He smiled and pointed at the stove. "With that and the heat from the machines, we often leave the door open at forty below," he said. I am shown a long silver tube as the visit comes to an end. It is a single-shot photographic device that goes down the hole and takes a picture of the azimuth and dips to see that the line of drill rods are on track.

There is time to drop in on one more rig. Fred, the Polish-born driller, does not seem surprised when two visitors open the door of his twelve-by-sixteen-foot drill shack. Drillers, I am told, have seen everything. Outside the door there is a series of boxes collecting slurry from the drill. The muck that settles in them will be analyzed for mineral content. My guide says the company is extremely careful about preserving the environment. He remarks as we drive back along the ice road about a time when he was on a job and oil cans fell off a transporter into a river. Management man or not, he ran through the water to pick them up. This land has to be left as it was found.

I talked about the job with Bob Franz later. He emigrated to Canada in the fifties, in search of a better life. In this case he found it in diamond drilling. Family photo albums are like a travel guide full of exotic locations with strange sounding names. There is one constant in all the pictures. The inevitable drill rig pokes skyward and the drill reaches down into the ground. Bob would not see himself as a salesman, but he speaks well of the industry. "It is not a soft job, drilling, but it pays well and there's never a dull moment," he says. Safety is very much his concern, but the odd story of troublesome skunks, bee stings, the rare plane crash and other misfortunes filter into the conversation. The job abounds with stories. Take Fred, the driller I met earlier. He was on a job on Baffin Island and radioed home. "I need a new engine," he said tersely. There was a pause. "What parts do you need?" queried the Kirkland Lake base. Another pause. "Make it another engine," came the reply. "But, man, why a new engine? asked the exasperated base man. "The engine we flew up was in good shape." Another burst of static on the receiver and then, "I need one because the helicopter let go of this one at five hundred feet!" A replacement was sent north pronto.

Next time you see a drill rig off on the skyline, remember the drillers. They will be running that drill day in and day out until the customer's requested depth is reached, no matter what it takes.

Scoop tram at Quebec mine exits the portal. The view is practically the same as at Kirkland Lake.

– LAC

LAC

Nobody goes to such depths to help Kirkland Lake.
Who says history doesn't repeat itself?

– Advertising slogans

LAC Minerals Limited will be one of the largest gold producers in Canada as this book goes to print. The firm grew from a group of small, widely dispersed companies to preeminence in its field due to many factors, most of which related directly to human resources. LAC deserves to have its complete story told one day. This brief account is not about LAC's gas and oil interests or mines in various locations in North America, including those in Quebec, where the Doyon Mine is probably the largest open-pit gold operation on the continent. Nor is this the place to touch on its potential at the great Hemlo camp. Instead see the company's efforts at Kirkland Lake as an old mining camp is revitalized.

Harry Rutetski is senior vice-president in charge of operations. He is a soft-spoken man who once worked underground as a production miner. He has, at the Preston East Dome, East Malartic and Manitouwadge, a total of twenty-five years with LAC. From his office overlooking the new Lakeshore Mine, Mr. Rutetski has furnishings a little different from those found in the offices of most senior executives of major Canadian corporations. There is a collection of historic miners' lamps. Ore samples cover desks, along with maps and hard hats.

I was free to visit the various plants and will pass on some impressions of what it is like in a mining operation for those who have never been in a mine.

Under its former name, Little Long Lac, the company acquired several of the former Kirkland Lake mines in the late fifties and also took over the Macassa Mine. As the climate for gold became more favourable, LAC embarked on an ambitious programme to sink a new shaft at Macassa, which at 7,275 feet will be the deepest single lift in North America, and also to reopen the once-great Lakeshore Mine. Of the accounts which follow, the trip underground at Macassa was taken in 1978, although the kinds of things seen would be the same today, while other other locations described were observed in December 1985.

Macassa Mine

The mine parking lot is almost full when I arrive at 6:45 a.m. The headframe looms against the blueing sky and except for the lights from the buildings, the rest of the mine seems deserted. The nearest building to the headframe is the combination office, battery room and dry. Through the door I see hard-hatted miners wait their turn to go underground. In a room ringed by longitudinal maps of workings, I receive a hard hat, battery-operated light and sign a release absolving the mine of responsibility in the event of injury.

There are safety signs everywhere, both from the Ministry of Mines and Macassa itself. Amid the background noise and flash of lights every time men move their hard hats, some facts emerge about the mine. The average tonnage is 247 tons per day for a total of 90,000 tons in a year. There are 285 employees and about 82 underground at one time on either of the two shifts. The mine manager studies reports of tonnage, locations of working places and a myriad of problems encountered during the last two shifts. A chain of command is quite evident here, but all conversation seems to run on a first-name basis. I overhead in one corner a studious discussion on the relative merits of black and red grease pencils, so necessary for marking rock underground.

Clothes hang from the roof in the Lakeshore Mine 'dry', 1935. — Author coln.

The "dry" is a big room with wash basins, showers and toilets. Chains with small baskets attached hang from the ceiling. As old as Canadian mining, the idea of a dry is both practical and simple. Miners change into work clothes, place their street attire on hangers attached to the chain and their valuables in the basket. Both are hauled to the ceiling and remain safe there until the end of the shift. The man responsible for the dry keeps it spotless. Next door, in the battery room, the compact four-volt batteries stand ready under charge. The battery tender tells me the light is good for eight hours. When charged the red-jacketed batteries automatically shut down from the charge and wait for the next shift. As we leave, my guide transfers two discs from one board to another. "That's in case we leave you behind!" he says with a grin. "If there are any discs on the left board at the end of shift, someone will form a search party."

The headframe is like a great hollow skyscraper. The cage tender is warmly bundled up from the chill, but apart from long underwear to soak up the damp and act as insulation, the group about to descend are not as heavily dressed. In the Kirkland Lake camp the temperature rises one degree for every 163.4 feet of depth, so there is no fear of catching cold underground. While we wait for the cage, someone points to piles of lumber in the shaft house and remarks that mines eat wood. Now the tender opens a wooden gate and about eight of us squeeze into a tiny steel box open to the sky. The cage has no roof and we are packed inside with mine surveyors and their equipment. There is a sampler off to take specimens of rock in specified areas. The tender signals the hoistman by jerking a cord to produce a series of staccato blasts. The hoistman sits in a separate building, the signals and marks on his big cable drum count as eyes in the business of moving human and other cargo. The cage starts off and gains momentum as we jolt down into the darkness, gathering speed to cover three fifths of a mile in well under a minute.

Stations or levels flash by at 150-foot levels like stations on a subway. The cage stops and the rope signal is used to notify the surface hoistman of our arrival. Up goes the bar, the steel doors of the cage clang open, the wooden doors swing wide and we walk to the next ride. Safety messages abound here as on surface and the visitor becomes aware of the importance of an ordered sequence of events. A winze, or underground hoist, drops us another 2,000 feet and the tour begins.

The newcomer has no hint of claustrophobia. The tunnels — they call them crosscuts to the ore and drifts along it — are six feet wide and seven feet high in this mine and often wider where ore cars meet. Air and water pipes run along the roof and the battery lights are perfectly adequate for vision. At intervals large fans force fresh air along the passageway. The floor is covered with a brownish mud. My guide informs me that pumps operate continually below the 4,000-foot level, keeping under control the water-filled now-dormant Kirkland Lake mines. A great deal of time is spent walking to places underground. Occasionally we pass men, but the impression is obtained of rather solitary work. There is no factory atmosphere here. The men we meet are of the multinational north country mix. All greetings come with a joke or a cheerful quip. I remark on the friendly manner and find that it is considered part of the job. Underground men depend on each other and there is no room for a grouch.

The sights and sounds encountered on that four-hour hike through the Macassa cannot all be recaptured, but a few impressions stand out. Processes in the mine work on the gravity principle. Ore, the rock containing the mineral, drops through ore passes. These are seen through gratings along the drift. By this means ore is concentrated in areas where it can be brought by railcars to the shaft for hoisting. At the end of one crosscut we come across the rock face. A miner operates a slusher, an air-powered car which lifts loose rock and deposits it behind on the vehicle. The guide points out the vein and the novice sees, at best, tiny yellow specks glinting in the rock. The machine is noisy as the whoosh of compressed air jerks it forward and back.

We visit various stopes — the term used to describe the excavation where ore is drilled, blasted and dropped through ore passes to waiting cars. Entry and exit to stopes involves much clambering up and down manways or ladders in vertical tunnels in the rock in this mine. Miners work in pairs, and when one needs the other, the cry "partner" echoes through the working place. In one location a miner is filling drill holes with blasting powder, another is setting up timber to support walls, while yet another is mucking the ore. A motorman is driving an ore train and we stand in a hollowed-out lay-by waiting for the machines to pass. My guide is a supervisor and notes the unusual and problems. A loose bar on the cage, fallen rock, shortage of material on one level, all will receive attention.

The cage glides back to surface and a swallow permits the ears to adjust to pressure changes. In the dry a shower reveals the amount of sweat and dust accumulated underground. As I leave, an impression remains of purposeful activity, and the wit and ingenuity of man so necessary to wrestle the gold from the rock.

Through the Old Macassa Mill

A visit to a gold mill is too much to take in for the novice, hence the diagram or flow chart. Picture a long, shed-like building with a blend of conveyor belts, tanks of various sizes and a maze of catwalks and stairways. Add to it a continual rumble of machinery and a rough picture of a mill emerges.

Before entering the mill, the ore has been crushed into small chips less than three eighths of an inch in diameter. Ball mills — huge revolving cylinders filled with iron balls — crush the now tiny rock to fine particles. Add water, lime and cyanide in solution and gold dissolves out of the rock paste into the solution. Then the mix is stirred and separated. Once the worthless rock is removed, that grey muddy substance that has passed through the system to this point is also removed and the solution or liquor is clear, just like water. Using zinc, the gold is precipitated onto canvas sheets in a press. Each week the valuable mud is scraped off the press, which looks something like a long, narrow radiator. With fluxing chemicals, unwanted metals are dissolved and a button of gold or silver is formed at the bottom of the furnace. This product is then melted down and poured into a mould to make a forty-pound bar, which is shipped to the Royal Mint in Ottawa.

The mill operates nonstop year round, for the process requires continual movement. At roughly half an ounce recovered for every ton of rock, there is a great deal of material to go out of the mill as waste, slimes or tailings. Even at this stage the tailings are checked to see if any gold is present. On the rare chance that this might happen, the tailings are passed through the mill circuit again. During the day twenty-four men, part of two twelve-hour shifts, run the mill. Jobs vary but include operators, mechanics and electricians.

There is only one disappointment for the visitor. The whole process is not very romantic. Milled gold is so fine in solution that not a bit was seen on the whole trip.

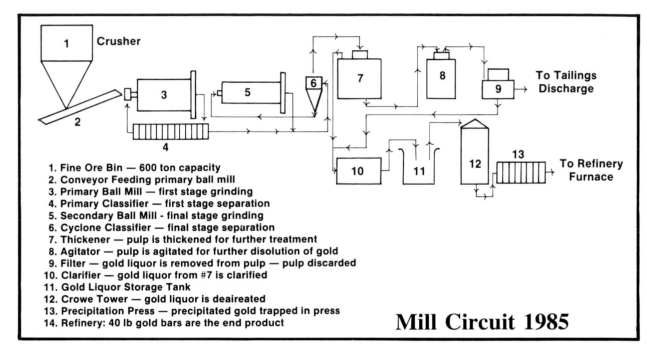

1. Fine Ore Bin — 600 ton capacity
2. Conveyor Feeding primary ball mill
3. Primary Ball Mill — first stage grinding
4. Primary Classifier — first stage separation
5. Secondary Ball Mill - final stage grinding
6. Cyclone Classifier — final stage separation
7. Thickener — pulp is thickened for further treatment
8. Agitator — pulp is agitated for further disolution of gold
9. Filter — gold liquor is removed from pulp — pulp discarded
10. Clarifier — gold liquor from #7 is clarified
11. Gold Liquor Storage Tank
12. Crowe Tower — gold liquor is deaireated
13. Precipitation Press — precipitated gold trapped in press
14. Refinery: 40 lb gold bars are the end product

Mill Circuit 1985

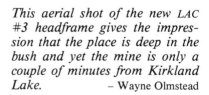

This aerial shot of the new LAC #3 headframe gives the impression that the place is deep in the bush and yet the mine is only a couple of minutes from Kirkland Lake.
— Wayne Olmstead

Shaft Sinkers

At the time I visited LAC's number three shaft at the Macassa Mine in 1985, the company was employing a shaft sinking firm, Dynatec, to take its new shaft down to 7,275 feet, a record for a single lift on the continent. I picked up a hard hat and yellow rain slicker for the drive to the site. My guide indicated that the project would come to $32 million when complete. The company felt the new facility would provide better working conditions, improved haulage and hoisting of ore. In addition, whereas now it took men almost an hour to get to their working place via hoist, underground winze and a long hike, the same distance could be covered through the deep hoist in about ten minutes.

Another safety message appeared. Get Rid of Loose. The meaning is lost on the uninitiated, but translated it means that all loose rock must be scaled or knocked down from over head to prevent accidents. The great shaft house looms up and next to it a dry is being constructed. Mines use contractors for such work so they can concentrate on mining, the job they know best. The shaft house is ringed by contractors' trailers. There are bright new aluminum skips for carrying ore, ready to go underground when the shaft is finished. Long piles of foot-square lumber are piled around the perimeter. Equipment waiting to be repaired for the shaft sinkers includes an air drill mounted on tracks. It looks like a small bulldozer but will drill up to fifty feet when stressing rock. Nearby lies a giant grab. Once in working order, the grab jaws are invaluable in the shaft, as it can pull up to three tons of rock at a time.

Inside the shaft house all is quiet, with no indication of the activity in the shaft below. Giant circuit breakers about eighty cubic feet in volume wait their turn to go underground. A separate compressor room now services the other Macassa shaft as well as this one. Giant underground pumps are monitored here. There is a huge room filled with electronics somewhat like computers in appearance. Behind them is a refrigeration room where air is cooled for underground. A small tunnel leads directly to the new shaft. All service cables run through it for ease of maintenance. At one corner a locked steel cage contains expensive drill bits of various sizes. Two giant drums in tandem hold the one-and-three-quarter-inch-thick hoist cable. The cable is inspected regularly for flaws.

Occasionally the shaft doors swing outward and a man climbs off the bucket, which serves as workhorse in the shaft until it is complete. Shaft sinkers wear all the usual safety gear and yellow waterproof coats and pants. Such men are the cream of the crop. They work in what could be extremely dangerous conditions and do everything as a team. Safety has to be their watchword if they want to earn their bonus. I did not see anyone chewing on his nails, but none of the shaft sinkers looked at all like timid men.

The hoist room is a glass-covered box complete with air conditioning for summer operation. The interior is like something out of a space movie, with flashing lights, dials and controls. The hoistman has to be very careful, as there are men working in the shaft just fifteen feet below the bucket which transports men and broken rock to surface. He defines his job as requiring a cool head, powerful concentration and a strong memory. When asked he can recite the progress of the shaft to the inch. His hoist can lift men 1,400 feet per minute while shaft sinking is in progress, or waste rock at 2,100 feet per sixty second span. The buzzer sounds and the hoistman manipulates small toggles which control the great hoist and raise the bucket. There is a safety device that takes over if the hoistman loses control. I am assured gravely that this will never happen.

After a tour of the new headframe and the preparations being made for regular operation, one is struck by the smooth multi-disciplinary approach to planning the whole venture.

Lakeshore Comes Alive

The great Lakeshore Mine produced 8.5 million ounces of gold in its forty-seven years. LAC reasoned that there must be more ore and even obtained maps showing reserves blocked out in various sections of the mine. There was also the crown pillar. This is the rock left as an umbrella to provide a roof for a mine. In the case of the Lakeshore, the crown pillar was extremely rich in ore. The first inclination was to get at it via open pit mining, but this hit a snag as the mine was covered with the slimes-filled Kirkland Lake directly above. Eventually it was decided to get at the ore by spiralling down via a ramp from surface. This is in the vicinity of the old incline shaft, which was formerly used to take timber and other supplies underground. An added bonus was the still serviceable main shaft of the mine. Two and a half million dollars later there was a new headframe in place, surface buildings were made ready for use and the inclined roadway was taking large vehicles.

I donned underground togs and walked with my guide from the portal, or entrance, to the incline. It seemed odd walking underground, but through a series of broad curves the trip to the 200-foot level was covered in a matter of minutes. In traditional mining, men go underground so as to remove ore from above. Here the gold ore is mined as found. Concrete is trucked in to support and seal working places. Rock mechanics are most important in an old mine with the possibility of unstable ground. The science of keeping rock in place is a serious occupation at the historic property. As usual safety signs predominate. One said, Remember Where Your Escape Routes Are.

In the crown pillar section the underground "road" is broad. A large truck passes carrying ore. My guide remarks that when this area is mined out, the work will revert to the two-man partner principle which works so well. We pass from the new workings to those which were in use when Harry Oakes owned the place. At one point the walk takes us by a shaft which is in perfect shape, yet no cages will ever travel it again. A modern shopping mall is sitting on top in the location of the former Lakeshore mill.

The Lakeshore Mine was the product of a good company. The mine was well-kept. It is often hard to distinguish between work done in the start-up period and that done in former years when the mine was in operation. The great number five shaft, which is serviced by the new hoist, was put down forty years before. The whole shaft is lined with fireproof concrete slabs which look as good as the day they were put in place. The cage tender who spots us from level to level is the same as all the others, a "mine" of information and a born comedian.

The trip comes in impressions. The safety man seems to follow us around. In one spot he puts in cables and bolts so that men working near the shaft can anchor themselves when necessary. In another area men are making slow going cutting through a blocked drift. My guide stops to question a miner getting ready to drill close to the shaft. This seems dangerous, but it turns out to be quite safe, as the small excavation will take only a fire door. Supplies are moved about with ore cars. In another area, two men are having lunch. They pause long enough to enquire about the weather on top. Back to surface, but this time in the cage. I notice the hoistman and decide he is the only one not moving in the whole place. One thing about mining, the times goes fast.

In four brief glimpses of LAC, the company which is revitalizing an old gold camp, one impression dominates. There is an enthusiasm and confidence which pervades all levels of the work force. Peter Allen, the president, talks about success due to the relationship between geological and production departments. Expand that and say that LAC runs well on the good relations between its people.

Visit the Mining Camps

This section is for people who wish to come north and explore the now familiar places where the great gold and silver rushes took place. Consider it also an invitation for previous visitors to perhaps look at the country from a different viewpoint. Only four locations are described, but the visitor will find much to see in between.

Mattawa

Mattawa is not one of the mining camps, but it is the place where Noah and Henry Timmins lived at the time they met with Fred Larose and their later historic connections with Cobalt and the Porcupine. This pretty town at the junction of the Ottawa and Mattawa rivers is located on Highway 17 about forty minutes' drive east of North Bay on the way to Ottawa. At first glance one is impressed by the fine nineteenth-century homes and interesting old stores on Main Street.

The Chamber of Commerce building at the corner of the Trans Canada Highway and the main shopping district has pamphlets outlining walking tours of Mattawa. At this point consider the people who came through this historic settlement. The place means "Meeting of the Waters" in the Iroquois language. Only thirty-nine miles to the north is Lake Temiskaming. The Iroquois meaning is apt, for the broad sweep of the river is very pleasing in all seasons. In 1611 Etienne Brule was the first white man to pass this way. Samuel de Champlain ended his first voyage of exploration here. A provincial park nine miles west honours the great explorer and offers displays of trade canoes and the voyageurs.

Mattawa was a centre of the fur trade for 200 years. Explorers, missionaries and fur traders passed through on their way to the West. This was a link in the canoe route with Lake Nipissing, the French River and Georgian Bay. The Jesuits came this way and several were martyred later in the Huron country. People like Radisson and Grosseilliers, the La Verendyres and Alexander Mackenzie saw the meeting place of waters and rested here on the long cross-country route.

Down Main Street, before you come to the Mattawa River road bridge, you see a building on the left-hand side of the street marked "Ike Tongue's old store." Tongue was a later owner, but this was originally the Timmins store. It was here that Fred Larose made the acquaintance of Noah Timmins and showed him his silver samples. That large building on the corner of Main and Gorman streets is much changed now. The high, peaked roof slopes back to the rear of the store in a series of steps. The side windows are blocked up, but the visitor can see the fancy brickwork on the front and scrollwork near the roof.

Across the road and down Timmins Street is the site of the original Timmins family home. The house was demolished in 1935 to make way for a park, and in 1938 the facility was dedicated by Noah's grandson, Gilles Timmins. Back across the bridge, visit Explorer's Point and the Mattawa District Museum, a fine new log building. West on Graham Street is Moosehead Lodge on Lake Chant Plein. Built by Henry Timmins, it was originally called "While-Away Camp." The place went through several owners and is now used as a Christian training centre.

Before leaving, visit the CPR station west on the way out of town. The building is much the same as Noah knew it — complete with the high, pitched roof — but the VIA Rail sign does nothing but detract from the place.

Cobalt

Go north ninety miles on Highway 11 from North Bay to Cobalt. Pass through the great bluffs overlooking the city. Highway 11 is the old Ferguson Highway, now a splendid but still sparsely used road with plenty of room and frequent passing lanes. North through the nature reserve the huge pines give some indication of what this country must have looked like before J.R. Booth's loggers cut it over. There is a tangle of rocks, trees and lakes all the way. Take 11B, the secondary road which leads to Cobalt, just a short distance north of Latchford. It is well signposted and refers to itself grandly as the Scenic Route. There is some truth to this claim, for the visitor gets a chance to slow down and look at the early days of Ontario's mining past. Be really prepared to drive slowly. The road twists and turns, and when it reaches Cobalt, the best speed is slow.

Come into Cobalt via Nickle Street, then right on Grandview and left on Silver Street. The Cobalt Mining Museum is probably the best place to start your exploration of the great silver camp. This account cannot do justice to Cobalt. It has an atmosphere all of its own and has to be savoured by the visitor. A guide to the town may be purchased at the museum, so this account will just touch on some of the highlights. Just inside the door is the mayor's ceremonial chain. It is of solid silver nuggets and is probably the most unusual in Canada. Close by is a picture of the donor, Mayor Lang, wearing his badge of office. The seven-room museum has a wide array of artifacts, not only from Cobalt but also Kirkland Lake and Porcupine. Note the elegant silver and tableware from the Cobalt Mess. There is a fine display of ore specimens, fluorescent rocks and old photographs.

– Cobalt Museum

One of the most unusual mayoral chains in Canada, this one of heavy leaf silver was donated by Mayor Lang in the boom days.
– Author

Just opposite the museum is "the biggest claim post in the world." Actually Cobalt streets are marked with claim posts, an interesting reminder of how the town began. There are also various mining relics placed at different locations in the town centre area. You cannot miss them. Just down the street, in a small park on the corner of Prospect and Silver streets, there is a display with the names of all the Cobalt mines. Marvel at the sheer number of them. Kitty corner across the street is the much-photographed headframe with the former store built around it. Long since disused, it once provided refrigeration. Food was just lowered down the shaft. Just up Prospect, a little behind the old Fraser Hotel, is the monument to Willet Miller. It reminds us he was the first provincial geologist of Ontario, from 1902 to 1925. In fine poetic form it says he "read the secrets of the rocks in New Ontario." Back down at the foot of Prospect is the famous Cobalt Square. Many buildings are changed now due to progress and numerous fires. The "square" does not have that shape at all, by the way. The elegant, now-disused station is at the foot of the square. Try to imagine the hustle and bustle here in the early days, with hundreds of silver bars waiting for shipment, untended on station carts.

The walking tour offers much to see, but for a change, work your way up a series of streets that cheerfully declare themselves to be dead ends and survey Cobalt Lake from one of the high spots. That water that you see on the lake bed is just on the surface. The slimes lie underneath. At one end there is a ball field. Across the lake see the Agnico Eagle property, once the Silver Miller and before that the famed Larose Mine. There are great open cuts in the cliffs where veins were followed until they petered out. Below St. Therese School, see the Right-of-Way Mine. The place is covered with rusty metal sheets now, but once the property which straddles the railways tracks was a great producer.

Go down to the lake area and cross the bridge by the Right-of-Way Mine. Take a look at the Larose cabin and curve right on the way to the Ragged Chutes. Stop off at the Nipissing lookout on the site of the camp's largest mine and view the slimes, great pits, trenches and scarified land surface. Many areas are fenced in now for visitor safety. Marvel at what was done here to change the face of the land in the all-consuming desire to recover silver ore. There are great concrete foundations everywhere. Some were for mills and for the other buildings about which one can only speculate.

On the way to view the compressed air plant at Ragged Chutes, take in the famed silver sidewalk at the Lawson Mine, now just a cut in the ground where men once scuffed their boots in the silver to keep it shiny. See also the Drummond cairn set in the chimney of the habitant poet and mine owner's former home. The track is better walked here, as it is rather rough. The road to Ragged Chutes offers many views of old mines and opportunities to explore, but few are named. Take a look at the remaining buildings and ore dumps. Look for blue-grey rock. Sometimes it is the bloom of cobalt ore, an oxidized powder with purple shading. In some places you may see a new operation at the site of one of the old mines. Perhaps the former producer is being explored again for silver missed in past years.

Hound Chutes is still a working hydro plant. Go on from here to Ragged Chutes. It is well worth the trip. Along the side of the road, the great air pipes snake over the contours of the land with huge concrete anchors placed along the route. At Ragged Chutes the rapids tumble over the rocks and remind us of when this enterprise provided Cobalt industry with compressed air. The section that was ruined in the fire is easily seen. One regrets that there is no opportunity to observe the spectacular geyser of water as the plant blows off, but hopefully one day the plant will be revitalized and put into operation once more.

Drive north on 11B through North Cobalt to Haileybury. Instead of turning down toward the lake, continue over the main road and go a couple of blocks to the Haileybury School of Mines. There are fine exhibits here of mining memorabilia, photographs of industry personalities and a good model of a mine. As you leave, take a look at the sculpture of a miner set in bas relief against the brick exterior of the college. This is one of six different sculptures donated to Northern College by the *Globe and Mail* when its former plant was torn down. Since William H. Wright owned the newspaper and was a resident of Haileybury for some years, it makes a fitting memorial to this great prospector and mine owner.

Drive down to the lake now. Practically all that you see was destroyed in the 1922 fire. Just behind the modern municipal building is the former fire station. The Fire Museum is located in this building and it is an excellent "hands on" small museum. Another spot worth visiting on the way out of town is the Haileybury Public Library. The visit is worthwhile, for the library houses the Temiskaming Art Gallery and there are always displays, many with northern themes. Just past the library is the scenic route along the lake. This section has been known for years as Millionaire's Row. There are fine, gracious houses, some of which were owned by mining and lumber magnates. Pass through the pleasant country town of New Liskeard and rejoin Highway 11. From here Kirkland Lake is only fifty miles. The Englehart and District Museum, on the way, is well worth a visit for those who have the time. Turn off the main highway onto number 112 and then east on number 66 to Kirkland Lake, the town that stands on gold.

Kirkland Lake

Enter Kirkland Lake through a large rock cut. In this vicinity the local name was once the Golden Gate, as it was the main entrance from the west to the camp. Goldthorpe Road on the left leads to the Macassa Mine, one of two gold mines operating in the town in the year this book was published. Pass well-kept houses on the mine property and stop to view the mine itself. Entry except on business is discouraged, but by all means take a photograph. Recall this was the last of the seven mines of the camp to be established. Turn back to the highway and continue east. Just opposite Thompson Honda is the Kirkland Lake Gold — latterly Minerals — property. The headframe of the mine Dr. Tyrell proved is gone now but some buildings remain. Down Beaver Hill, see on the right the skeleton of the south Teck Hughes Mine headframe. The north headframe is long since gone to another location, in Quebec. This was the place where Sandy McIntyre had his second chance before his earnings trickled through his fingers.

The Sir Harry Oakes Chateau is well signposted on the left. A parking lot is provided and a short walk brings the visitor to Harry's great log house, built in the Western Prairie school of architecture. Pamphlets about the building as well as a brief article on Oakes, along with books and a variety of souvenirs, may be obtained inside. The main floor is now devoted to exhibitions, but a bust of Harry himself greets the visitor on entry. The second floor has rooms with displays of pioneers, northern history, rocks and animals. Be sure to see Nancy Oakes' bedroom. It presents a stuccoed surface but the plasterer was an artist. Examine the walls carefully and notice familiar nursery rhyme figures worked into the stucco. Turn away, and while one view is obscured, another comes into view. The top floor is closed except to researchers. In its original form as the attic, the area was so large as to allow family square dances there. Leave the second floor by the back of the house and take an opportunity to gaze over what was Kirkland Lake. On leaving the museum, go round to the side where once the Oakes' cars were garaged. Take another look at Kirkland Lake from this point.

Just a little further along the road is a modern shopping mall. It is on the site of the former giant Lakeshore mill. Across the road see the Don Lou Motel. This was once the Lakeshore bunkhouse where single men lived very well. The building is much changed, but take a look at the original copper roof. Turn down along the side of the mall and see a neat beige, metal-covered headframe. LAC Minerals has spent much time and money rehabilitating the Lakeshore Mine. The present headframe is small compared to the former one, but the visitor can see the operation from a short distance and take photographs. You may

Gold towns are not shy about advertising their products. Kirkland Lake's Mayor Joe Mavrinac wears his gold badge of office.
– Town of Kirkland Lake

see very ordinary dump trucks leave the mine. They are carrying gold ore to the mill up at the Macassa Mine.

Kirkland Lake's main street, Government Road, is a switchback of curves and rock. It has been improved considerably over the years. See the building next to Dino's Pizzeria. This was the location until recently of the Ash Can, more commonly known as the Kirkland Lake Hotel. On the left you can see Kresges, where Roza Brown once lived. Down farther on the left is Charlie Chow's Hotel. Now turn left on Prospect Avenue, pass the I.G.A. store and the Commodore Motel, and you are now on the former Wright-Hargreaves Mine property. Heath & Sherwood, the drilling firm mentioned earlier, is located down the road on the right. Notice a small park. Stop and see the monument to all the town hockey players who made their way to the National Hockey League. Just behind the monument are the mine gates, moved from their original location but kindly brought back to Kirkland Lake by the Hargreaves family. The liquor store is situated right on the former mine shaft. Old-timers say all the bottles will fall down it one day.

Go back to Government Road and continue east. The school on your right is built on top of the former Townsite Mine. A few hundred yards to the left there is a hydro station. Behind this building was the Sylvanite property, but it was so well demolished that practically nothing remains. Not too much further on the main road is the elegant Toburn headframe. Pass it and go up to see Northern College. It stands on the old Tough-Oakes property and has that mine's office vault on display with an historical plaque. Set in the walls of the college, see sculptures similar to the one at the Haileybury School of Mines. The visitor could spend a profitable day or two exploring Kirkland Lake. When finished drive back the way you came and continue to Swastika. By the river see another plaque and the ducks on the water. Many who made it rich and others who made nothing lived in this pleasant community in the old days. Just under the bridge see Culver Park coming up on the left. The Swastika Mine rock dump can still be explored on the right of the lake behind the new sewage treatment plant. Now get set for the ride to Porcupine.

Timmins and The Porcupine

The drive to the gold camp is about an hour and a half if you can avoid the various spots of interest along the way. Continue on Highway 66 to Highway 11 and head north. Pass through Kenogami and then settle down for a drive through bush country spotted with numerous lakes. At Matheson, drop by the museum located on the left-hand side of the highway. It contains a fine history of the area and has much on the great fire. From Matheson see a blend of farm land and bush, as the clay belt offers a more open prospect than the rocky country to the south. Leave the main highway and turn on Highway 101 to Timmins. Along this area reflect on the three-day hard packing in to the camp from the rail line; that was the price of a visit prior to 1911. Your car will do the job in about half an hour.

Just before Porcupine, see a huge building on the right. It is the concentrator of Kidd Creek Mines, formerly Texas Gulf. Just a short distance further is the Pamour Mine. In 1985 the company decided it required the land where the highway was located and so the road was changed to run south of the property rather than between the mine houses and the mine, as was formerly the case. Noranda owned Pamour and most of the formerly independent mines for several years, but recently control was purchased by an Australian firm. Those great hills just ahead are man-made. They are tailings from the mine, long since grassed over.

Porcupine is the small community which was once a tent city after the 1911 fire. In a small park by the lake see the vault of the first mining recorder in the camp. Note also one of the Hollinger locomotives and other smaller artifacts. Just across the road is a grassy area dedicated to George Bannerman. From here you can see the headframe of the Dome Mine and in the distance the mines of Timmins. Cross the causeway and remember that in the water to your left people sought refuge from the smoke and flames in 1911.

Go on past Northern College, through South Porcupine, and turn left at the traffic lights. Drive down to Bruce Street and see the Timmins Museum on your right. The museum has first-rate outdoor displays which are well marked. The building is superbly planned as a museum. One large area is for permanent displays and specialized collections. One display is called "Rainbow Chasers," what an apt description of the prospectors. Another area accommodates travelling exhibits. The gift shop offers a large selection of souvenirs and the museum has a full calendar of events.

The visitor could now continue on Bruce Street and return to Highway 101. An interesting optional trip which still winds up in Timmins is to turn left at the South Porcupine hospital and take the road past the great Dome Mine. This great mine has been operating since 1911. A newer headframe appears on the left-hand side. Further along see a headframe covered in blue metal. This mine was formerly the Preston East Dome. Like many area mines, this property has attractively laid out housing. Further down the road, on the right, see the old Porcupine Paymaster Mine and opposite it the Buffalo Ankerite property. Two more mines off to the left are the Aunor and Delnite.

Just past the golf course, turn right at the Gulf Oil bulk dealership and continue until you see the Hollinger park on the outskirts of Timmins downtown area. The huge mine is still, but on the former property Steetly Talc operates a fine-grinding plant. The Hollinger property has a well-maintained, modern headframe and shaft, and when the price of gold is right, that machinery will start operation again. In the meantime there is some open pit mining on the property.

Continue east on Highway 101, in the reverse direction to which you entered the Porcupine district. On the right, behind Feldman Lumber is the former Hollinger office building. The Chamber of Commerce at Schumacher is well worth a stop for area information. Obtain here particulars of the underground mine tour. The tour is highly popular and will likely have to be booked in advance.

Figure skating champion Barbara Ann Scott trained at the McIntyre Arena many years ago. The great McIntyre Mine has gone through some name changes, but that is the name which sticks with local people. A tiny headframe near Pearl Lake is the old Jupiter Mine. It is dwarfed by the McIntyre. Return to Highway 101 and see on the right another small park with mining implements displayed. Go left on the main road, and on the left a short distance away see the monument to Fred Schumacher, namesake of this community. The plaque reads '' . . . his memory will live forever.'' It will among children of Schumacher, who still receive Christmas toys from the estate of this generous community builder.

The tour ends here. Why not drive back into Timmins and explore the town? It is a vigorous, bustling community that proudly proclaims its mining background. The barber shops are full of ore samples. See the rock wall at the Seantor Motor Hotel. Many other stores and hotels have mining themes. Enjoy Timmins, the town founded on gold.

Phone booth formerly located next to the Chamber of Commerce building at Timmins, celebrates the gold rush by appearing as a head frame. – Author

The Porcupine Paymaster Mine. – Author

Gold and Silver

Gold

Gold is an element (Au). It comes in massive or thin plates, also in flattened grains or scales. Distinct crystals are very rare. There is no cleavage to the metal; it has a hardness of 2.5 to 3.0 on a scale to 10 and its specific gravity runs between 15.6 and 19.3. Gold comes in yellow colour and streak. It is extremely heavy, very malleable and ductile.

Silver

Silver is an element (Ag). It comes in flattened grains or scales. In rare forms it comes in wire-like shapes or irregular needle-like crystals. There is no cleavage; it has a hardness of 2.5 to 3.0 and its specific gravity runs from 10.0 to 11.0. The colour and streak are silvery white but may be tarnished grey or black. The element is highly malleable and ductile. There is a mirror-like lustre on untarnished surfaces.

The ancients called gold noble, as it was not changed in fire or oxidized like base metals. Silver was noble but less than gold, as it could be changed in aqua fortis or nitric acid.

One estimate gives all the gold mined in a 6,000-year period as only 80,000 tons. That is not much compared to many other minerals. Gold has influenced man because of its symbolic value. The yellow metal is so precious that every year refineries shut down simply to recover gold dust. Since it is relatively soft, gold is usually alloyed with copper or silver. Most native gold has some silver in it. Gold is measured in its fineness or carats, thus 22 carat is twenty-two parts gold and two parts copper. Another way of saying the same quality would be that it is 91.60 fine. One troy ounce of gold can be beaten into one hundred square feet of foil or drawn into a wire fifty miles long. Use it to plate a copper or silver wire and the end product can be up to 1,000 miles long.

The north abounds in stories of precious metals which are in place but . . . Consider Cobalt Public School. It is situated on the fine slimes of the once-great Coniagas Mine. In 1956 the school board prudently employed a mining engineer to discover if the public body owned something of value. Exhaustive tests showed that the property contained 11,000 tons, with a weighted average of 5.8 ounces of silver to the ton. At approximate values for today, the school is sitting on just over half a million dollars worth of silver. There is just one problem. When the test drilling was done it also brought up evidence of muskeg, logs and stumps. The conclusion in 1956 was that the tailings were worthless because there was no way of extracting the silver because of the debris underneath.

A Mining Primer

Some of the terms in this book require a little explanation for readers unfamiliar with mining. Not all technical words from the text are included, some are explained elsewhere.

Assay: Testing ores or minerals to determine the amount of valuable metals contained.

Bullion: Precious metal in bars or a form other than coins.

Cage: Steel compartment used to transport men and equipment vertically in a shaft.

Collar: Timbering or concrete around the mouth of a shaft.

Crosscut: A horizontal opening or underground tunnel driven from a shaft to an ore structure.

Crown Pillar: A solid block of ore left near the surface of a mine to provide a roof.

Cut and Fill: A mining method in which ore is removed in slices or lifts. The excavation is then filled with waste material before the next slice is taken.

Drift: A horizontal underground tunnel which follows the vein or tunnel, as opposed to a crosscut which intersects it.

Face: In a drift, crosscut or stope, the end where work is progressing.

Geology: The science of the earth's crust, its strata and their relation and changes.

Geophysics: Scientific method of prospecting that detects minerals by utilizing their physical properties. These include magnetism, specific gravity, electrical conductivity and radioactivity.

Grubstake: Finances, food or supplies advanced to a prospector in return for a share in any discoveries made.

Hanging Wall: The wall or rock on the upper or top side of a vein or ore deposit.

Headframe: The framework on surface above a shaft which supports the hoisting cables. When sheeted in, it is the shaft house.

High-grade: Rich ore. A highgrader is one who steals ore.

Hoist: The machine for raising and lowering the cage.

Level: The horizontal passageways on a mine working horizon. Levels are usually at fixed distances apart and measure from surface, perhaps 150 foot. 300 foot, etc.

Manway: Passage for men only to travel via stairs or ladder, often as an escape route in case of emergency.

Mineral: A naturally occurring substance with definite physical properties and chemical composition, having crystal form in some cases.

Muck: Ore or rock which has been broken up by blasting. The name gives no indication as to value.

Ore: A mixture of minerals from which at least one of the metals can be extracted at a profit.

Prospect: A mining property, the value of which has yet to be proved.

Quartz: A kind of mineral, massive or crystalline in hexagonal prisms.

Raise: A vertical or inclined underground working excavated from the bottom up.

Rock: Any naturally formed combination of minerals forming part of the earth's crust.

Sedimentary Rocks: These are laid down underwater and come from other rock particles. They are usually laid in layers.

Shaft: The vertical or inclined excavation which opens and services a mine. There is usually a hoist to raise and lower men and materials.

Sheave Wheel: Large grooved wheel at the top of a headframe, over which the hoisting cable passes.

Shrinkage Stope: Means of stoping (see below) which uses part of the broken ore as a working platform as well as support for the walls.

Skip: A large self-dumping type of bucket used in a shaft to hoist ore or rock.

Station: Larger area where a level meets a shaft; used mainly for storage and equipment handling. It is just like a station on a subway, hence the name.

Stope: Any working place in a mine where ore is or has been extracted.

Tailings: Material rejected from a mill after the valuable minerals have been removed. Tailings are often in the form of a sand-like substance and are sometimes referred to as "slimes."

Vein: A fissure or crack in a rock filled by rocks that have travelled upwards from deep in the earth.

Winze: This is similar to a shaft, but whereas a shaft starts at surface, this is an opening sunk from below ground.

Working in the stope.   – Author coln.

Songs of the Mining Camps

The Cobalt Song

You may talk about your cities and all the towns you know,
With trolley cars and pavements hard and theatres where you go,
You can have your little auto and carriages so fine,
But it's hobnail boots and a flannel shirt in Cobalt town for mine.

CHORUS: For we'll sing a little song of Cobalt,
If you don't live there it's your fault.
Oh, you Cobalt, where the big gin rickies flow,
Where all the silver comes from,
And you live a life and then some,
Oh, you Cobalt, you're the best old town I know.

Old Porcupine is a muskeg, Elk Lake a fire trap,
New Liskeard's just a country town and Haileybury's just come back;
You can buy the whole of Latchford for a nickel or a dime,
But it's hobnail boots and a flannel shirt in Cobalt town for mine.

We've got the only Lang Street: there's blind pigs everywhere,
Old Cobalt Lake's a dirty place, there's mud all over the square,
We've got the darnedest railroad, that never runs on time,
But it's hobnail boots and a flannel shirt in Cobalt town for mine.

We've bet our dough on hockey and swore till the air was blue,
The Cobalt stocks have emptied our socks with the dividends cut in two,
They didn't get any of our money in darned old Porcupine,
But it's hobnail boots and a flannel shirt in Cobalt town for mine.

The author acknowledges writer L.F. Steenman, his estate, and Cobalt Kiwanis Club.
"The Cobalt Song" was written in February 1910.

The Porcupine Song

Silver and gold, in this country cold are sought by each of us.
When old Larose once stubbed his toes, it made quite a lot of fuss.
But wait until the next summer, when the sun begins to shine,
I'll show you a mine, in dear old Porcupine,
Where the gold is rich and fine, and I hope that I get mine!

CHORUS: For I have warts on my fingers, and corns on my toes,
Claims up in Porcupine, Goodness only knows!
So put on your snowshoes, and hit the trail with me
To P-o-r-c-u-p-i-n-e! That's me!

Over the snow went "Right-of-Way" Joe, to see this find so grand.
The gold in the quartz was so big as warts upon a schoolboy's hand.
Along with him went A.A. Cole, that four-eyed engineer.
Said Joe: "Now have no fear. We know the stuff is here.
To Cobalt we will steer, and we'll sing this song so queer . . ."

CHORUS: For I have warts . . .

S.R. Heakes had heard for weeks about this Porcupine;
At last said he: "Let's go and see if it is just as fine
As what they say." So he took his way as far as Pearl Lake.
Said he: "This is no fake! I guess I'd better stake,
And then my claims I'll make another Kerr Lake!"

CHORUS: For I have warts . . .

Perhaps you think it's easy to get to Porcupine;
You hire a rig; you think you're big; you start out feeling fine.
When you get to Father Paradis; you eat some pork and bean.
You don't feel very lean; you think you're nice and clean;
But you scratch and cuss and scream, and it ain't the pork and bean.

CHORUS: For YOU have warts . . .

Warts on your fingers, corns on your toes,
Claims up in Porcupine, CRUMBS in your clothes,
So, put on your SNOWPACKS, and hit the mud with me
To P-O-R-C-U-P-I-N-E! That's me!

The author acknowledges writers Eddie Holland, Jack Leckie, Scotty Wilson and their estates. The words were written in 1910 and 1911.

Kirkland Lake Song

There's a town a-way up north of Cobalt
That is world famous for its gold.
The spirit there is just the same, I'll tell you,
As when men discovered it in days of old.

CHORUS: Well! What do you know, we're from Kirkland Lake,
    We're the boys and we're the girls who'll always give and take;
    It's a real town, a gold town, we're in it with a stake.
    We come from near, we come from far, to good old Kirkland Lake.
    Now we think a lot of Cobalt and Porcupine is great,
    But here's to the Pride of the whole wide North;
    Here's to Kirkland Lake.

The author acknowledges writer Jack Reid and his estate.

Selected Bibliography

Baldwin, D., and Dann, J.A. *Pictorial History of Silver Mining*. Cobalt: Highway, 1976.

Barnes, M. *Gold Camp Pioneer — Roza Brown of Kirkland Lake*. Cobalt: Highway, 1973.

_____. *Link With A Lonely Land*. Erin: The Boston Mills Press, 1985.

_____. *The Town that Stands on Gold*. Cobalt: Highway, 1978.

Bocca, G. *The Life and Death of Sir Harry Oakes*. London: Weidenfield & Nicolson, 1959.

Boucher, M. *Our Temiskaming*. Cobalt: Highway, 1976.

Cassidy, G. *Arrow North*. Cobalt: Highway, 1976.

Epps, E., and Bray, M., eds. *Vast and Magnificent Land*. Toronto: Lakehead and Thunder Bay Universities, 1984.

Fancy, P. *Silver Centre*. Cobalt: Highway, 1985.

Gard, A. *Echoes From Silverland*. Toronto: Emerson, 1909.

_____. *Gateway To Silverland*. Toronto: Emerson, 1909.

_____. *Silverland*. Toronto: Emerson, 1907.

Gibson, T.W. *Mining In Ontario*. Toronto: King's Printer, 1937.

Girdwood, C., Jones, L., and Lonn, G. *The Big Dome*. Toronto: Cybergraphics, 1983.

Gilluly, J., Water, A., and Woodford, A. *Principles of Geology*. San Francisco: Freeman, 1959.

Gough, B. *Gold Rush*. Toronto: Grolier, 1983.

The Discoverers. M. Hanula, ed. Toronto: Pitt, 1983.

Helm, N. *Tri-Town Trolleys*. Cobalt: Highway, 1984.

Higley, D. *The History of the Ontario Provincial Police*. Toronto: Queen's Printer, 1984.

Hoffman, A. *Free Gold*. New York: Associated Book Services, 1946.

Hogan, B. *Cobalt — The Year of the Strike 1919*. Cobalt: Highway, 1979.

Innes, H. *Settlement and the Mining Frontier*. Toronto: Macmillan, 1935.

Leasor, J. *Who Killed Sir Harry Oakes?* London: Heineman, 1983.

LeBourdais, D. *Metals and Men*. Toronto: McClelland and Stewart, 1957.

Historical Highlights of Canadian Mining, ed. R. Longo. Toronto: Pitt, 1973.

Lonn, G. *About Mines and Men*. Toronto: Pitt, 1964.

_____. *The Mine Finders*. Toronto: Pitt, 1964.

Lougheed, W. *The Gold Community*. Toronto: Timmins Industrial Commission, 1958.

MacDougall, J.B. *Two Thousand Miles of Gold*. Toronto: McClelland and Stewart, 1946.

McDowell, L.S. *Remember Kirkland Lake — The Gold Miners' Strike of 1941-2*. Toronto: The State and Economic Life Series, University of Toronto Press, 1983.

McRae, J. *Call Me Tomorrow*. Toronto: Ryerson, 1960.

Murphy, J. *Yankee Takeover at Cobalt*. Cobalt: Highway, 1977.

Nellis, H.V. *The Politics of Development*. Toronto: Macmillan, 1974.

Newman, P. *Flame of Power*. Toronto: Longmans, 1959.

Newton, W. *Westward with the Prince of Wales*. Toronto: McLeod, 1920.

Pain, S. *The Way North*. Toronto: Ryerson, 1960.

_____. *Three Miles of Gold*. Toronto: Ryerson, 1966.

Phillips, L. *Noranda*. Toronto: Clarke Irwin, 1956.

Peterson, O. *The Land of Moosoneek*. Timmins: Anglican Diocese of Moosonee, 1974.

Bob Miner and Union Organizing, ed. W. Roberts. Hamilton: Labour Studies Programme, McMaster University, 1979.

Robinson, A. *Gold in Canada*. Ottawa: King's Printer, 1933.

Randles, A.C. "The Geology of the Porcupine Range." PhD. Thesis, University of Toronto, 1951.

Schull, J. *Ontario Since 1867*. Toronto: McClelland and Stewart, 1972.

Stortroen, M. *Immigrant in Porcupine*. Cobalt: Highway, 1977.

Stovel, J. *A Mining Trail*. Kingston, Queen's University, 1906.

Townsley, B.F. *The Mine Finders*. Toronto: Saturday Night Press, 1935.

Tucker, A. *Steam Into Wilderness*. Toronto: Fitzhenry and Whiteside, 1978.

Williamson, O. *The Northland*. Toronto: Ryerson, 1946.

Young, S., and Young, S. *O'Brien*. Toronto: Ryerson, 1967.

Pamphlets, Papers and Articles

"The Davis Handbook of the Cobalt Silver District." *Canadian Mining Journal*, 1910.

Timmins, N.A. "A Reminiscent History." *The Canadian Mining Journal*, September 1935.

"The Porcupine Mining District." *The Canadian Mining Journal*, Porcupine series II, 1911.

Trethewey, W.G. "Early Days in Cobalt." *The Canadian Mining Journal*, January 1909.

The Canadian Annual Review, 1906.

The Canadian Mines Handbook 1985-1986.
    Toronto: Northern Miner Press.

The Cobalt Daily Nugget.

Munro, A.E. *An Era of Progress — The Draining of Cobalt Lake and Its Effect on the Town*. Cobalt: The Daily Nugget Press, 1913.

Silver and Gold. Cobalt: The Daily Nugget Press, 1916.

Brown, L.C. "Cobalt Blooms Again." *Canadian Geographical Journal*, July 1953.

————. "Cobalt — The Town with the Silver Lining." *Canadian Geographical Journal*, July 1963.

"Cobalt-Silver-Nickel-Arsenic Ores of the Temiscamingue District." Canadian Mining Institute transactions, 1905.

"The LAC Minerals Story." *Canadian Mining Journal*, May 1985.

Knight, C.W. "Prospecting in Ontario." *Canadian Mining Journal*, Vol. 71, 1951.

Kee. H.A. "Sinking Operations at McIntyre #11 Shaft." *Canadian Mining and Metallurgical Bulletin*, April 1926.

Who's Who in Canadian Mining — Ontario and Quebec. Timmins: Directories North, 1985.

Brown, L.C. *Golden Porcupine*. Toronto: Queen's Printer, 1974.

The Porcupine Gold Fields and the Cobalt Silver Mines. Toronto: Grand Trunk Railway, 1911.

Annual Reports 1982-1985. Toronto: LAC Minerals.

The Northern Daily News, Kirkland Lake.

"The Florence Mining Company versus The Cobalt Lake Mining Company." Toronto: The Ontario Court of Appeal, 1909.

The Ontario and Quebec Goldfields. Legislative Library of Ontario, 1911.

The Porcupine Advance, Timmins.

The Story of the Prospector and the Porcupine 1909-1939. Timmins: The Porcupine Prospectors' Association, 1939.

Report of the Royal Commission on Mineral Resources of Ontario. Toronto: Queen's Printer, 1981.

Stopps, W. *A Walk Through Cobalt's Past.* Cobalt: Cobalt Miners' Festival, 1978.

The Timmins Daily Press, Timmins.

Cole, A.A. *The Mining Industry in that Part of Ontario Served by the Temiskaming and Northern Ontario Railway.* Toronto: T. & N.O., 1913.

Acknowledgements

*The utmost care has been taken to guard against the
inclusion of puffs and inaccuracies.*

– The Davis Hand Book of the
Cobalt Silver District, 1910

A letter to *The Northern Miner* was one of the things that opened doors for this book. All sorts of interesting folks read the *Miner* and some responded with photographs and comments. Each contributor is gratefully acknowledged.

Mrs. A. Little put me in touch with George Caldbick's daughter. Bob Atkinson, the postcard king, was most helpful. Ken Docherty of the Timmins Museum aided in a photo search, as did Carolyn O'Neill at the Sir Harry Oakes Museum. Retired O.P.P. Superintendent Jim Tappenden put me on the trail of a police picture.

Gordon Peacock not only loaned pictures but gave me some gold samples and made me understand the feeling prospectors have for rocks.

Desmond Woods of Boston Creek loaned photographs and mentioned the Bargnesi-Anastasia story. Peter Matijek confirmed the public facts.

Peter Ginn talked about the early days of Swastika and gave his father's account of Harry Oakes' real reason for leaving Canada. Mrs. Joan Craig, Ed Hargreaves daughter, graciously gave me some background on her father's partnership with her uncle, Bill Wright.

Ron Purdy of Cobalt gave me a copy of the report on the value of the tailings in his schoolyard. Hans de Ruiter told me his version of the Saint Barbara story and displayed his statue of the miners' patron.

At LAC Minerals, Harry Rutetski, senior vice-president operations, Paul Holmes, manager of the Lakeshore Division, Andy Kuchar, general superintendent, explained the operations and arranged for visits to the properties. If the section on LAC seems simple, it is because underground mining is a mystery to many people and the idea was to present the work in a straightforward fashion.

Rick Regimbal, then contract manager at Heath & Sherwood, took me through the diamond drill business, both in the office and in the field, with a fine dose of humour. Bob Franz, field supervisor, kindly filled in any gaps.

Joyce Allick at the Teck Centennial Library and Brian Cahill and Jonathan Lee at Ontario Library Service, James Bay, gave their usual sterling service. For the unknowns at the National Library, great thanks for the way they track down books and other items.

Elderhostel students visiting Northern College in Kirkland Lake are always interested in northern mining and sparked the idea for this book.

My wife Joan is the one to whom I owe the most. She forgives an oft-absent writer-researcher husband.

In any book where research forms the basis for the work, the author always has the fear that there will be unintentional errors, that in some way copyright may be breached or some kind person who has aided in some way has not been given credit. I hope once again that none of these things have happened.